Virus Infections and the Developing Nervous System

Virus Infections and the Developing Nervous System

Edited by

Professor R. T. Johnson

Eisenhower Professor of Neurology
Professor of Microbiology and Neuroscience
The Johns Hopkins Hospital
Baltimore
USA

and

Professor G. Lyon

Service de Neurologie Pédiatrique
Laboratoire de Neurologie du Développement
Université Catholique de Louvain
Cliniques Universitaires Saint-Luc
Brussels
Belgium

KLUWER ACADEMIC PUBLISHERS
DORDRECHT / BOSTON / LONDON

Distributors

for the United States and Canada: Kluwer Academic Publishers, PO Box 358, Accord Station, Hingham, MA 02018-0358, USA
for all other countries: Kluwer Academic Publishers Group, Distribution Center, PO Box 322, 3300 AH Dordrecht, The Netherlands

British Library Cataloguing in Publication Data

Virus infections and the developing nervous system.
  1. Young animals. Nervous system. Pathogens
Viruses
  I. Johnson, R.T.   II. Lyon, Gilles
  591.2'188

  ISBN-13: 978-94-010-7051-5

Library of Congress Cataloging in Publication Data

Virus infections and the developing nervous system.

  Includes bibliographies and index.
  1. Central nervous system—Infections. 2. Virus
diseases in children. 3. Central nervous system—
Infections—Animal models. 4. Developmental neurology.
I. Lyon, Gilles.   II. Johnson, Richard Tidball,
1931–
RJ496.I53V57   1988   618.92'804194   88-6817
ISBN-13: 978-94-010-7051-5   e-ISBN-13: 978-94-009-1243-4
DOI: 10.1007/978-94-009-1243-4

Copyright

Contents

CONTENTS

List of Contributors

O. BOESPFLUG
Laboratoire de Neurovirologie
INSERM U 56
Hopital de Bicetre
94275 Le Kremlin Bicetre Cedex
France

A. BURNY
Department of Molecular Biology
University of Brussels
67, rue des Chevaux
1640 Rhode-St-Genèse
Belgium

Y. CLEUTER
Department of Molecular Biology
University of Brussels
67, rue des Chevaux
1640 Rhode-St-Genèse
Belgium

L. G. EPSTEIN
Departments of Neurosciences and
    Pediatrics
UMD–New Jersey Medical School
185 South Orange Avenue
Newark
NJ 07103
USA

B. N. FIELDS
Department of Microbiology and Medical
    Genetics
Harvard Medical School
25 Shattuck Street
Boston
MA 02115
USA

B FLECKENSTEIN
Institut für Klinische und Molekulare
    Virologie
Loschgestrasse 7
8520 Erlangen
FRG

C. GODFRAIND
Laboratoire de Neurologie du
    Developpement
Université Catholique de Louvain
Cliniques Universitaires St Luc
Brussels 1200
Belgium

R. T. JOHNSON
Department of Neurology
The Johns Hopkins University School of
    Medicine
Baltimore
MD 20205
USA

P. G. E. KENNEDY
Department of Neurology
Institute of Neurological Sciences
Southern General Hospital
Glasgow G51 4TF
UK

R. KETTMANN
Faculty of Agronomy
5800 Gembloux
Belgium

K. KRISTENSSON
Department of Pathology (Neuropathology)
Karolinska Institute
Huddinge Hospital
Huddinge
S-141 86 Sweden

F. LEHMANN-GRUBE
Heinrich-Pette-Institut für Experimentelle
    Virologie an der Universität Hamburg
Martinstrasse 52
2000 Hamburg 20
FRG

J. LÖHLER
Heinrich-Pette-Institut für Experimentelle
  Virologie an der Universität Hamburg
Martinstrasse 52
2000 Hamburg 20
FRG

A. LÖVE
Department of Virology
Karolinska Institute
SBL
Stockholm
S-105 21 Sweden

M. MACH
Institut für Klinische und Molekulare
  Virologie
Loschgestrasse 7
8520 Erlangen
FRG

M. MAMMERICKX
National Institute for Veterinary Research
99 Groeselenberg
1180 Uccle
Belgium

G. MARBAIX
Department of Molecular Biology
University of Brussels
67, rue des Chevaux
1640 Rhode-St-Genèse
Belgium

E. NORRBY
Department of Virology
Karolinska Institute
SBL
Stockholm
S-105 21 Sweden

D. PORTETELLE
Faculty of Agronomy
5800 Gembloux
Belgium

G. A. SCANGOS
Molecular Therapeutics, Inc.
400 Morgan Lane
West Haven
CT 06516
USA

J. A. SMALL
Department of Neurology
Johns Hopkins University School of
  Medicine
Baltimore
MD 21205
USA

M. TARDIEU
Laboratoire de Neurovirologie
INSERM U 56
Hopital de Bicetre
94275 Le Kremlin Bicetre Cedex
France

R. THOMAS
Department of Molecular Biology
University of Brussels
67, rue des Chevaux
1640 Rhode-St-Genèse
Belgium

B. D. TRAPP
Department of Neurology
Johns Hopkins University School of
  Medicine
Baltimore
MD 21205
USA

K. L. TYLER
Department of Microbiology and Medical
  Genetics
Harvard Medical School
25 Shattuck Street
Boston
MA 02115
USA

U. UTZ
Institut für Klinische und Molekulare
  Virologie
Loschgestrasse 7
8520 Erlangen
FRG

A. VAN DEN BROEKE
Department of Molecular Biology
University of Brussels
67, rue des Chevaux
1640 Rhode-St-Genèse
Belgium

L. WILLEMS
Faculty of Agronomy
5800 Gembloux
Belgium

J. S. WOLINSKY
Department of Neurology
The University of Texas Health Science
  Center at Houston
PO Box 20708
Houston
TX 77225
USA

1
The Viral Infections of the Developing Nervous System: An Overview

R. T. JOHNSON

A landmark in clinical observation occurred in the summer of 1941 after Norman Gregg, an Australian ophthalmologist, saw 20 infants with congenital cataracts. He obtained information from colleagues on 47 other infants with cataracts and found that two-thirds also had microcephaly and over half had congenital heart disease and low birth weights. Most important, he obtained a history that all of the mothers had had German measles in the early months of pregnancy during a preceding winter epidemic. He postulated that this might represent 'an infectious process resulting in a partial arrest of development'[1].

In 1956 both Rowe and his colleagues[2] and Smith[3] isolated cytomegalovirus from human tissues. This virus was quickly associated with the cytomegalic inclusion disease, which had been recognized as a cause of 1% of infant deaths, and was subsequently associated with frequent inapparent infections. However, 20% of these 'silent infections' proved to result in subsequent hearing loss or mental retardation[4, 5]. These findings established cytomegalovirus as the commonest known cause of mental retardation.

Herpes simplex virus had been associated with neonatal infections for many years, but in the late 1960s the two forms of herpes simplex were differentiated. The type 2, or genital herpes, was found to cause most infections of neonates and to cause widespread encephalitis often with death or severe sequelae[6]. Very rarely herpes simplex has been found to cause congenital microcephaly with infection during gestation[7]. Similarly, on rare occasions, varicella during pregnancy can result in a cicatricial scarring and limb hypoplasia in infants[8]. Coxsackie B virus infection in the neonatal or in late gestational period has occasionally led to encephalitis; poliovirus and arbovirus infections in late gestation have led to congenital poliomyelitis or encephalitis[9]. However, rubella, cytomegalovirus and herpes simplex remained as the three major infections of the developing human nervous

1

Table 1.1 Viral infections of the nervous system in the human fetus or neonate

Virus	Time of infection	Neurological disease
Major importance		
Rubella	First trimester	Chronic encephalitis, microcephaly, retardation, diplegia, visual and auditory deficits
	Later gestation	Hearing loss and minor retardation
Cytomegalovirus	During gestation	Silent infection with minor mental and auditory deficits; cytomegalic inclusion disease with microcephaly, retardation, motor, visual and auditory deficits
Human immunodeficiency virus	Unknown – probably during gestation and neonatal period	Chronic encephalopathy with acquired microcephaly, diplegia, and basal ganglia calcification
Herpes simplex virus	During parturition	Severe encephalitis
Minor importance		
Herpes simplex virus	During gestation	Encephalitis with congenital microcephaly
Varicella zoster virus	First half of gestation	Cicatricial scar, limb hypoplasia, and encephalomyelitis
Coxsackie, group B virus	Neonatal (?late gestation)	Encephalitis associated with myocarditis
Polioviruses	Late gestation	Congenital or neonatal paralytic poliomyelitis
Arboviruses	Late gestation	Congenital or neonatal encephalitis
Influenza virus	First trimester	Doubtful relation to malformations
Lymphocytic choriomeningitis virus	?	Hydrocephalus (?)
BK virus	?	Hydrocephalus, microcephaly (?)

Modified from Johnson, R. T. (1982)[9].

system until the past three years. Human immunodeficiency virus (see Chapter 5) must now be added as a fourth major cause of fetal or neonatal infection that leads to severe neurological deficits (Table 1.1).

THE PATHOGENESIS OF FETAL INFECTIONS

Infection in the mother is a prerequisite to fetal infection, but often this infection is clinically inapparent. Virus can reach the fetus by infection of the germinal cell or genetic transmission of the viral genome as seen in lymphocytic choriomeningitis virus in mice[10] or with murine leukemia viruses[11], respectively. Alternatively, virus can cross the placental barrier and enter the blood or amniotic fluid, and infect the fetus as occurs with rubella and cytomegalovirus infections in man. Finally, there may be ascending

Table 1.2 Possible effects of viral infections on the developing nervous system

---

Indirect (maternal infection only)
    Constitutional effect on mother
    Infection of placenta

Direct fetal infection (acute or chronic)
    Generalized encephalitis with necrotic or inflammatory lesions
    Chromosomal damage
    Mitotic inhibition
    Infection of selected cell populations altering embryogenesis

---

Taken from Johnson, R. T. (1982)[9].

infection or infection within the birth canal as in type 2 herpes simplex virus infections.

Fetal damage or fetal wasting can result from constitutional effects of a systemic infection on the mother or because of infection limited to the placenta (Table 1.2). The high rate of abortion that occurred during smallpox epidemics was thought to be associated with the systemic effects on the mother. Years ago, Mims and I inoculated ectromelia virus, a poxvirus of mice, into pregnant mice with resultant abortion without fetal infection. There was, however, widespread infection of the placenta. Similarly, mouse cytomegalovirus infects the placenta during clinically inapparent infections of mothers and causes fetal death[12], unlike the human cytomegalovirus that does infect the fetus. When a virus does gain access to the fetal germinal cells, fetal circulation or amniotic fluid, the effects on the fetus are dependent on the developmental stage, on which cell population is susceptible to infection, and on what effect the viral infection has on those cells. A cytolytic infection will cause destructive lesions; but a non-cytopathic infection, such as lymphocytic choriomeningitis virus in mice, may cause persistent infection of cells throughout the body including neurons of the brain without obvious ill effects.

Viruses are also capable of producing chromosomal damage but there is little evidence that this occurs in humans. Mitotic inhibition is also reported in cell culture systems and the small number of cells and organs of children with rubella, their small birth weight despite gestational age, and the finding of viral antigen in normal cells in malformed organs all suggest that there may be multifocal infection with mitotic inhibition occurring with this disease[13]. Persistent non-cytopathic rubella virus infection of human embryonic mesenchymal cells depresses their response to epidermal growth factor[14]. Finally, viral infections of selected cell populations can alter embryogenesis and lead to symmetrical non-inflammatory deficits that lack any pathological features to suggest an antecedent infection[15]. These abnormalities resemble malformations previously thought to have a genetic basis. Infection can alter organogenesis, selectively destroy particular immature cell populations or disrupt differentiated cells in immature animals where the repair process leads to a histological appearance of agenesis. Examples of these forms of malformations are known only in animals but may well have counterparts in human disease (Table 1.3).

Table 1.3 Experimental cerebral malformations induced by viruses

Malformation	Virus	Host species
Defects in neural tube closure	Myxoviruses: Influenza virus Newcastle disease virus	Chick embryo Chick embryo
Encephalocoele	Togavirus: St. Louis encephalitis virus	Mouse
Cerebellar hypoplasia	Parvoviruses: Rat virus Feline panleukopenia virus Minute virus of mice	Hamster, rat, cat, ferret Cat, ferret Mouse
	Arenaviruses: Lymphocytic choriomeningitis virus Tamiami virus	Rat Mouse
	Togaviruses: Hog cholera virus Bovine viral diarrhoea virus Border disease virus	Pig Cow, sheep Sheep
Hypomyelination	Togaviruses: Hog cholera virus Bovine viral diarrhoea virus Border disease virus	Pig Sheep Sheep
Hydranencephaly and/or porencephaly	Bunyaviruses: Akabane virus Rift Valley fever virus	Cow, sheep Sheep
	Orbiviruses: Bluetongue virus	Cow, sheep
	Togaviruses: Venezuelan equine encephalitis Wesselsbron virus Bovine viral diarrhoea virus Border disease virus	Monkey Sheep Sheep Sheep
Hydrocephalus[a]	Poxvirus: Vaccinia virus	Cat
	Paramyxoviruses: Mumps virus Parainfluenza virus types 1, 2, 3 Canine parainfluenza Respiratory syncytia virus Pneumonia virus of mice Measles (mutant) virus Newcastle disease virus	Hamster, monkey Hamster, mouse Dogs Hamster, mouse Mouse Hamster Mouse
	Orthomyxoviruses: Influenza virus	Hamster, mouse, monkey
	Reoviruses: Type 1	Mouse, hamster, rat, ferret
	Togaviruses: Ross River virus St. Louis encephalitis virus	Mouse Mouse

[a] Includes communicating and non-communicating hydrocephalus but obstructive hydrocephalus due to virus-induced neoplasia and hydrocephalus *ex vacuo* due to chronic infection are not included[16].

EXPERIMENTAL VIRAL INFECTIONS ON THE DEVELOPING NERVOUS SYSTEM

Neural tube defects

The first experimental study of the tetragenic effect of virus on the nervous system was in 1947 by Hamberger and Habel[17]. They inoculated influenza A virus into 48-hour-old chick embryos and induced abnormalities of the neural tube. After 24 h of further incubation, the embryos had poor development of the primitive brain and abnormalities of tube flexion. Subsequent histological studies showed failure of closure of the neural tube but this was not associated with any cytopathology in the neural ectoderm, notochord or surrounding mesenchymal tissue. Mitotic activity along the luminal sides of the tube was normal[18]. We repeated these studies with immunofluorescence staining to determine which cells were infected and were surprised to find infection limited to chorionic and amnionic membranes and the non-neural ectoderm[19]. In a similar study with Newcastle virus causing the same malformation in chick embryos immunofluorescent studies had shown infection of cells of the caudal neural tube[20]. Therefore, the same teratogenic effects resulted when different cells were infected by different viruses. The nature of the defect appeared to be dependent on the ontogenic state when the insult occurred rather than on a specific target cells of the infection.

Hydranencephaly and porencephaly

Hydranencephaly and porencephaly were first found as sequelae to viral infection in 1955. This occurred under curious circumstances. Bluetongue virus, an arthropod-borne orbivirus, causes respiratory and gastrointestinal disease in sheep but not neurological disease. After the disease was recognized in the United States, a live attenuated virus was developed by serial passage of virus in chick embryos, and it was field tested as a vaccine in 1955. The attenuated virus produced no clinical disease, induced antibody, and immunized sheep resisted challenge. During the subsequent lambing season, however, a large number of ewes immunized during pregnancy delivered lambs with malformed brains[21].

Silverstein and his coworkers were injecting this virus directly into fetal animals to study the retinal dysplasia that it produced[22]. During their experiments they found consistent cerebral malformations, and in our subsequent collaborations we found that fetuses inoculated at 50 days of their 150 days of gestational period developed hydranencephaly and those inoculated at 75 days of gestation developed porencephaly. Inoculation at 100 days of gestation produced no gross abnormalities and only microscopic nodules of the brain[23]. Immunofluorescence studies showed a selective infection of the subventricular zone of germinal cells of the forebrain[24]. Early infection of these cells led to widespread necrosis and cavitation of the cerebral hemispheres; whereas infection later in gestation led to infection largely in glial cell precursors with smaller areas of necrosis leading to cavitation largely in the white matter. However, by the time of birth, inflammatory cells and

5

necrotic tissue were removed leaving a smooth glial walled cavity with no viral antigen, inflammatory cells, or other footprint of the antecedent infectious process.

Granuloprival cerebellar degeneration

A selective infection of cerebellar germinal cells leading to malformations was reported in 1964 by Kilham and Margolis[25]. Kilham had isolated a rat parvovirus that after inoculation into newborn hamsters caused a selective destruction of the external germinal cells of the cerebellum prior to their postnatal migration to form the granular layer. This led to a cerebellum with an abnormal foliation, total or subtotal absence of a granular cell layer, and an abnormal synaptic organization[26]. This experimentally induced malformation bore a remarkable resemblance to a disease in cats known as spontaneous ataxia of kittens, the commonest neurological disease in domestic cats, regarded for 100 years as an autosomal recessive degenerative disorder. Indeed, they found that it was caused by feline panleukopenia virus, a parvovirus[27]. In adult cats this virus infects mitotic cells in the bone marrow and intestinal wall, and in the late gestational neonatal kitten the mitotic germinal cells are destroyed. In the fetal kitten, bone marrow and intestinal cells are replaced prior to birth but the germinal neurons cannot be replenished. The kittens at birth appear normal, because neonates are normally ataxic, but as they fail to achieve smooth motor control the deficit becomes obvious giving the illusion of a progressive degenerative disease. Furthermore, the pathology shows only the absence of a specific cell layer without residual evidence of inflammation or infection[26]. The similarity of these lesions to granuloprival cerebellar generation in man is startling[28]. The recent recognition of human parvoviruses[29] raises the possibility of a similar mechanism in the human disease.

Aqueductal stenosis and hydrocephalus

My initial interest in the effects of virus infection on the developing nervous system arose from serendipitous findings while trying to devise an experimental model for post-infectious encephalomyelitis. In 1964, I began studies of a neurotropic strain of mumps virus that caused a fatal disease in neonatal hamsters characterized by intense perivascular inflammatory responses[30]. Subsequent studies, however, showed that a non-cytopathogenic infection of neurons killed the animals[31], and the search for an animal model of post-infectious encephalomyelitis was unsuccessful. During the course of those studies, however, 'control' animals were inoculated with a strain of mumps virus recently isolated from the cerebrospinal fluid of a child with aseptic meningitis. This wild-type strain of mumps virus did not cause any acute clinical disease in suckling hamsters. However, mumps virus was found to grow but growth was limited to ependymal cells. Three to six weeks later, these animals began losing weight and showing signs of cranial enlargement. Hydrocephalus was found secondary to stenosis or occlusion of the aqueduct of Sylvius[32]. By the time that clinical disease developed the inflammatory reaction had resolved. Small clusters of ependymal cells persisted in the area of stenosis forming rosettes or aqueductules and the area normally occupied by the aqueduct was not replaced by glial tissue but contained normal brainstem parenchyma[33]. Thus, at the time of clinical disease or death, virus antigen and inflammation had cleared and the stenotic aqueduct fulfilled all of the histological criteria for primary agenesis of the aqueduct[34]. Subsequently, a variety of other viruses have been shown selectively to infect ependyma and have been shown to cause similar lesions. This can even occur with defective infection as in influenza virus infection in mice[35].

Subsequent to the animal studies a number of children have been reported with hydrocephalus and aqueductal stenosis 5 weeks to 4 years after mumps virus infection[16]. In electron microscopic studies of spinal fluid sediment of children with benign mumps meningitis, we found large numbers of ependymal cells often containing cytoplasmic tubules consistent with mumps virus nucleocapsids[36].

RELATING HUMAN MALFORMATIONS TO VIRAL INFECTIONS

Most pregnancies are associated with several viral infections but attempts to associate such infections with birth defects in the infant pose formidable problems[37]. The association of rubella virus, cytomegalovirus, and human immunodeficiency virus with birth defects has been relatively straightforward. Rubella is an epidemic disease with stereotypic clinical signs. The rate of malformations following maternal infection is high and the infected newborn shows rather characteristic signs. Hence the association of rubella with birth defects was possible as an astute clinical observation. Cytomegaloviruses were associated with birth defects because of a high rate of disease in newborns (representing 1% of the infant deaths), and characteristic pathological features with cytomegalic cells, inclusion bodies and inflammation suggesting a viral aetiology. Human immunodeficiency virus is causing epidemic disease, most mothers have been in a high risk group of intravenous drug users, and the children show characteristic signs in early childhood. Furthermore, each of these three viruses persists so that recovery of virus and identification of viral nucleic acids and viral antigens in lesions allows a solid linkage of the virus with the disease.

It has been possible to associate early intrauterine herpesvirus infections with rare malformations because of very characteristic syndromes. Most infants in whom intrauterine herpes simplex virus have been associated with microcephaly and calcification and atrophy of the cerebral hemispheres have had postnatal vesicular rashes from which the virus can be recovered. Infants in whom intrauterine varicella-zoster infections have been related to malformations have had mothers with clear-cut histories of chickenpox during pregnancy and a scar and hypoplasia in a uniquely zosteriform pattern.

From experimental studies we know that malformations can be caused by viruses when no virus can be recovered postnatally and no pathological features are found to give a clue to an antecedent infection. If a single virus produced rare malformations, that variable depending upon gestational age as found with the bluetongue vaccine infections in fetal sheep, or if a single malformation was an occasional result of many different viral infections as found with the hydrocephalus produced by many viruses, the association of the anomaly with antecedent maternal infection would defy present methodologies.

References

1. Gregg, N. M. (1941). Congenital cataract following German measles in mother. *Trans. Ophthalmol. Soc. Aust.*, **3**, 35–46
2. Rowe, W. P., Hartley, J. W., Waterman, S., Turner, H. C. and Heubner, R. L. (1956). Cytopathogenic agent resembling human salivary gland virus recovered from tissue cultures of human adenoids. *Proc. Soc. Exp. Biol. Med.*, **92**, 418–24
3. Smith, M. G. (1956). Propagation in tissue cultures of a cytopathogenic virus from human salivary gland virus (SGV) disease. *Proc. Soc. Exp. Biol. Med.*, **94**, 424–30
4. Reynolds, D. W., Stagno, S., Stubbs, G., Dahle, A. J., Livingston, M. M., Saxon, S. S. and Alford, C. A. (1974). Inapparent congenital cytomegalovirus infection with elevated cord IgM levels. Causal relation with auditory and mental deficiency. *N. Engl. J. Med.*, **290**, 291–6
5. Hanshaw, J. B., Scheiner, A. P., Moxley, A. W., Gaev, L. and Abel, V. (1975). CNS sequelae of congenital cytomegalovirus infection. In Krugman, S. and Gershon, A. A. (eds.) *Infections of the Fetus and Newborn Infants.* pp. 47–54. (New York: Alan R. Liss Inc.)
6. Nahmias, A. J., Alford, C. A. and Korones S. B. (1970). Infection of the newborn with herpesvirus hominis. *Adv. Pediatr.*, **17**, 185–226
7. Montgomery, J. R., Flanders, R. W. and Yow, M. D. (1973). Congenital anomalies and herpesvirus infection. *Am. J. Dis. Child.*, **126**, 364–6
8. Paryani, S. G. and Arvin, A. M. (1986). Intrauterine infection with varicella-zoster virus after maternal varicella. *N. Engl. J. Med.*, **314**, 1542–6
9. Johnson, R. T. (1982). *Viral Infections of the Nervous System.* (New York: Raven Press)
10. Mims, C. A. (1968). Pathogenesis of viral infections of the fetus. *Progr. Med. Virol.*, **10**, 194–237
11. Rowe, W. P. (1972). Studies of genetic transmission of murine leukemia virus by AKR mice. I. Crosses with Fr-1″ strains of mice. *J. Exp. Med.*, **136**, 1272–85
12. Johnson, K. P. (1969). Mouse cytomegalovirus: Placental infection. *J. Infect. Dis.*, **120**, 445–50
13. Rawls, W. E. and Melnick, J. L. (1966). Rubella virus carrier cultures derived from congenitally infected infants. *J. Exp. Med.*, **123**, 795–816
14. Yoneda, T., Urade, M., Sakuda, M. and Miyazaki, T. (1986). Altered growth, differentiation and responsiveness to epidermal growth factor of human embryonic mesenchymal cells of palate by persistent rubella virus infection. *J. Clin. Invest.*, **77**, 1613–21
15. Johnson, R. T. (1972). Effects of viral infections on the developing nervous system. *N. Engl. J. Med.*, **287**, 599–604
16. Johnson, R. T. (1975). Hydrocephalus and viral infections. *Dev. Med. Child Neurol.*, **17**, 807–16
17. Hamberger, V. and Habel, K. (1947). Teratogenetic and lethal effects of influenza A and mumps viruses on early chick embryos. *Proc. Soc. Exp. Biol. N.Y.*, **66**, 608–17
18. Robertson, G. G., Williamson, A. P. and Blattner, R. J. (1960). Origin of myeloschisis in chick embryos infected with influenza-A virus. *Yale J. Biol. Med.*, **32**, 449–63
19. Johnson, K. P., Klasnja, R. and Johnson, R. T. (1971). Neural tube defects of chick embryos: an indirect result of influenza-A virus infection. *J. Neuropathol. Exp. Neurol.*, **30**, 68–74
20. Williamson, A. P., Blattner, R. J. and Robertson, G. G. (1965). The relationship of viral antigen to virus-induced defects in chick embryos. Newcastle disease virus. *Dev. Biol.*, **12**, 498–519
21. Schultz, G. and DeLay, P. D. (1955). Losses in newborn lambs associated with bluetongue vaccination of pregnant ewes. *J. Am. Vet. Med. Assoc.*, **127**, 224–6

22. Silverstein, A. M., Parshall, C. J., Jr., Osburn, B. I. and Prendergast, R. A. (1971). An experimental, virus-induced retinal dysplasia in the fetal lamb. *Am. J. Ophthalmol.*, **72**, 22–34

23. Osburn, B. I., Silverstein, A. M., Prendergast, R. A., Johnson, R. T. and Parshall, C. J. (1971). Experimental viral-induced congenital encephalopathies. I. Pathology of hydranencephaly and porencephaly caused by bluetongue vaccine virus. *Lab. Invest.*, **25**, 197–205

24. Osburn, B. I., Johnson, R. T., Silverstein, A. M., Prendergast, R. A., Jochim, M. M. and Levy S. E. (1971). Experimental viral-induced congenital encephalopathies. II. The pathogenesis of bluetongue vaccine virus infection in fetal lambs. *Lab. Invest.*, **25**, 206–10

25. Kilham, L. and Margolis, G. (1964). Cerebellar ataxia in hamsters inoculated with rat virus. *Science*, **143**, 1047–8

26. Margolis, G. and Kilham, L. (1968). In pursuit of an ataxic hamster, or virus-induced cerebellar hypoplasia. In *International Academy of Pathology Monograph*. Vol. IX, pp. 157–83. (Baltimore: Williams and Wilkins Co.)

27. Kilham, L. and Margolis, G. (1966). Viral etiology of spontaneous ataxia of cats. *Am. J. Pathol.*, **48**, 991–1011

28. Sarnat, H. B. and Alcala, H. (1980). Human cerebellar hypoplasia. A syndrome of diverse causes. *Arch. Neurol.*, **37**, 300–5

29. Plummer, F. A., Hammond, G. W., Forward, K., Sekla, L., Thompson, L. M., Jones, S. E., Kidd, I. M. and Anderson, M. J. (1985). An erythema infectiosum-like illness caused by human parvovirus infection. *N. Engl. J. Med.*, **313**, 74–9

30. Kilham, L. and Overman, J. R. (1953). Natural pathogenicity of mumps virus for suckling hamsters on intracerebral inoculation. *J. Immunol.*, **70**, 147–51

31. Johnson, R. T. (1968). Mumps virus encephalitis in the hamster. Studies of the inflammatory response and noncytopathic infection of neurons. *J. Neuropathol. Exp. Neurol*, **27**, 80–95

32. Johnson, R. T., Johnson, K. P. and Edmonds, C. J. (1967). Virus-induced hydrocephalus: Development of aqueductal stenosis in hamsters after mumps infection. *Science*, **157**, 1066–7

33. Johnson, R. T. and Johnson, K. P. (1968). Hydrocephalus following viral infection: The pathology of aqueductal stenosis developing after experimental mumps virus infection. *J. Neuropathol. Exp. Neurol.*, **27**, 591–606

34. Russell, D. S. (1949). *Observations on the Pathology of Hydrocephalus.* (London: HMSO)

35. Johnson, R. T. and Johnson, K. P. (1969). Hydrocephalus as a sequela of experimental myxovirus infections. *Exp. Mol. Pathol.*, **10**, 68–80

36. Herndon, R. M., Johnson, R. T., Davis, L. E. and Descalzi, L. R. (1974). Ependymitis in mumps virus meningitis. Electron microscopical studies of cerebrospinal fluid. *Arch. Neurol.*, **30**, 475–9

37. Johnson, R. T. (1979). Problems in relating viral infections to malformations in man. *Contrib. Epidemiol. Biostat.*, **1**, 138–46

2
Reovirus Infections of the Central Nervous System

K. L. TYLER AND B. N. FIELDS

INTRODUCTION

Reoviruses are neurotropic viruses which have proven to be extremely useful in identifying genetic and molecular mechanisms of viral pathogenesis[1-5]. In this chapter we will focus on summarizing current knowledge concerning reovirus infections of the central nervous system (CNS), with particular emphasis on illustrating how studies with reoviruses have led to a better understanding of the pathogenesis of CNS infections.

ENTRY OF VIRUSES INTO THE HOST

Viruses which infect the CNS enter the host through a variety of routes including through the skin, the respiratory tract, the enteric tract and the genitourinary tract[6,7]. Studies with reoviruses[8,9] have provided insights into the precise cellular pathways by which enteric viruses penetrate the intestinal epithelium to enter the host via the intestinal route. Following oral inoculation into 10-day-old mice, reoviruses spread to Peyer's patches (focal aggregates of lymphoid tissue) and subsequently to mesenteric lymph nodes[10]. A virtually identical sequence of events occurs after feeding poliovirus to chimpanzees[11], suggesting that this may be a basic pathway for the entry of a number of enteric viruses into the host. Electron microscopic (EM) studies of the spread of reoviruses from closed ileal loops in mice indicate that virus initially adheres to the surface of specialized epithelial cells ('M cells') overlying Peyer's patches. Virus is then transported across these cells within intracellular vesicles and ultimately appears as free virus in the extracellular space of Peyer's patches.

The capacity of reoviruses to grow in the intestine appears to be a major determinant both of their subsequent capacity for systemic spread[12] and their subsequent transmissibility[13]. Genetic studies using reassortant reoviruses have indicated that proteins present on the outer surface (capsid)

11

of the virus, such as[12] the outer capsid protein μ1c and[13] the core spike protein λ2 are major determinants of these properties. Studies on the mechanisms of chemical and physical inactivation of reoviruses *in vitro* indicate that the outer capsid proteins are major determinants of the sensitivity of reoviruses to variations in pH and temperature[14,15]. This suggests the possibility that one mechanism by which outer capsid proteins influence rates of viral growth in the intestine may be through their effects on virus stability in the ambient environmental conditions present in the gut.

SPREAD OF VIRUS TO THE CNS

Once a neurotropic virus has successfully entered the host it must still spread to reach its ultimate target cells within the CNS. Classic studies using a variety of neurotropic viruses led to the recognition that the two principal pathways by which viruses reach the CNS are through the bloodstream ('haematogenous spread') and via nerves ('neural spread')[16–18]. It was concomitantly recognized that a variety of host factors including age, immunity, nutritional status and genetic background were important determinants of the capacity of viruses to spread via these routes[16]. More recently, reoviruses have been used to identify the role of specific viral factors in determining the capacity of a virus to spread via specific routes in the infected host.

As discussed earlier, after oral inoculation reoviruses can be sequentially detected in intestinal tissues, Peyer's patches and mesenteric lymph nodes. Reovirus type 1 Lang (T1), but not reovirus type 3 Dearing (T3), is also capable of spreading from mesenteric lymph nodes to the spleen, presumably by means of local lymphatic spread followed by bloodstream invasion. Reassortant viruses, containing various combinations of genes derived from T1 and T3, were used to show that the reovirus outer capsid protein σ1 determines the capacity of reoviruses to spread to extraintestinal organs such as the spleen following peroral inoculation[10].

The pathways by which reoviruses spread to the CNS after intramuscular inoculation in neonatal mice have been extensively investigated. In this model system, T3 uses neural spread to reach the spinal cord, and this spread can be completely inhibited by nerve section. Additional studies using selective inhibitors of fast (e.g. colchicine) and slow (e.g. β,β'-iminodiproprionitrile; IDPN) axonal transport indicate that neural spread of T3 is mediated via the microtubule-associated system of fast axonal transport[19]. This result is in accordance with earlier electron microscope studies in which T3 was visualized within axons[20] and dendrites of neutrons in close relationship to microtubules[21,22].

In distinction to T3, T1 spreads to the spinal cord via the bloodstream, and as a consequence this spread is not affected by nerve section or inhibitors of axoplasmic transport[19]. These results correlate well with previous studies which indicate that T1 generates a high titre viraemia after inoculation into the footpad, or intraperitoneally, in newborn mice[23–25]. Using reassortant reoviruses it is possible to show that the reovirus S1 double-stranded RNA segment ('gene'), which encodes the σ1 outer capsid protein, is responsible for determining whether these reassortant viruses spread to the spinal cord via neural or haematogenous routes after intramuscular inoculation[19].

The role of specific viral genes in determining the capacity of neurotropic viruses to spread via nerves has also been studied for herpesviruses. Unfortunately, the region of the genome that has been identified as important in spread encodes a large number of proteins (i.e. the viral DNA polymerase, the major nucleoprotein p40, the major DNA binding protein ICP8, and an envelope glycoprotein gB)[26,27]. Which of these proteins are actually important in determining the capacity of herpesvirus to spread to the CNS, and the mechanism by which this effect occurs, remains unknown.

VIRUS INVASIONS OF THE CNS

For viruses which spread to the CNS via nerves there is no specific anatomical barrier to invasion of the brain parenchyma. The problem is more complex for blood-borne viruses which must still penetrate the 'blood–brain barrier' to invade CNS tissue. Reoviruses may use several pathways to enter the CNS from the bloodstream. After subcutaneous inoculation T1 has been detected in choroid plexus epithelial cells [28]. This suggests that one potential pathway of CNS invasion for reoviruses may involve spread from blood into the stroma of the choroid plexus through the fenestrated endothelial cells lining choroid plexus capillaries. Virus could then infect the choroid plexus epithelial cells. Infection of choroid plexus epithelium provides an easy pathway for virus into the cerebrospinal fluid (CSF), and from the CSF to the ependymal cells lining the ventricles. A virtually identical pathway has been proposed for invasion of the CNS by viruses such as mumps[29].

A second potential route by which reoviruses could invade the CNS from the bloodstream is via capillary endothelial cells. T1 antigen has been detected in the endothelial cells of blood vessels throughout the body after subcutaneous inoculation and subsequent viraemia[25]. In one study, T3 (HEV strain) particles were seen inside brain capillary endothelial cells by EM after intraperitoneal inoculation of virus[20], although in another study T3 (Abney strain) could not be found in these cells[21]. These studies at least raise the possibility that reovirus viraemia may lead to infection of brain capillary endothelium with subsequent spread of virus into adjacent brain parenchyma. This route of CNS invasion may also be utilized by a number of other neurotropic viruses including many arboviruses[30].

Reoviruses can bind to lymphoid cells[31]. T3 viral particles have been visualized by EM within mononuclear cells and macrophages within the CNS[21,22], and inside lymphocytes within the lumen of brain capillaries[20]. In the brains of animals infected with reovirus, monocytes adhere to the inner lining of cerebral blood vessels and enter the CNS parenchyma following diapedesis across the endothelium of small cerebral veins[22]. These observations suggest the possibility that reoviruses could enter the CNS inside or in association with circulating lymphoid or mononuclear cells. A similar pathway of CNS invasion may be important for other neurotropic viruses, including lentiviruses[32] and parvoviruses[33].

VIRAL TROPISM IN THE CNS

Once invasion of the CNS has occurred, the pattern of clinical disease produced by a neurotropic virus appears to depend largely on the regions of the CNS affected, and the specific populations of cells within that region that are injured.

T1 and T3 produce different patterns of CNS infection in neonatal rodents. This difference has been exploited to identify the role of individual viral genes and the proteins they encode in determining the capacity of reoviruses to produce distinct patterns of neurologic injury. Following intracerebral, subcutaneous or peroral inoculation, T1 produces severe ependymitis. Cytoplasmic inclusion bodies appear in ependymal cells, which subsequently become necrotic and slough off into the ventricular cavities. Large contiguous regions of ventricle may become completely denuded of ependymal cells. Surviving cells proliferate and this proliferation, combined with sloughed cellular debris, can result in the production of non-communicating obstructive hydrocephalus. Blockage of the aqueduct of Sylvius appears to be a common sequela of T1 infection, and can result in massive dilation of the lateral and rostral IIIrd ventricles[28, 34].

Communicating hydrocephalus also appears to be common in T1 infections, and may result from blockage of CSF outflow into the basal cisterns or from interference with CSF readsorption at the arachnoid villi[35]. This type of hydrocephalus appears to be a consequence of an intense basilar meningitis, as well as inflammation and fibrosis involving the arachnoid villi.

The exact incidence and time of appearance of T1-induced hydrocephalus appears to be affected by both the dose and strain of virus used and the route of inoculation. In general, intracerebral inoculation of high doses of virus results in earlier appearance and higher incidence of hydrocephalus[35]. A number of additional neuropathological features occur consequent to the development of hydrocephalus including marked thinning of the cortical mantle, and caudal displacement of posterior fossa structures. Involvement of neurons is never a prominent feature of T1 infections[36, 37], although rare infected neurons have been seen in some studies[28, 35].

The pattern of CNS injury produced by T3 is completely distinct from that produced by T1. After inoculation via the intracerebral, subcutaneous, intramuscular or intraperitoneal routes, T3 produces a severe necrotizing encephalitis. Inclusions appear in the soma and dendrites of neurons throughout the CNS. The brunt of injury is typically most severe in the hippocampus and limbic system, and in the cortex (especially the cingulate gyrus and occipital lobes). Lesions are frequently seen in thalamic and hypothalamic nuclei, lateral geniculate body, anterior colliculus and cerebellar hemispheres[22, 38, 39]. Leptomeningitis is typically a prominent feature in T3 infections. Virus can be seen in glial cells and reactive gliosis can occur. Infection of ependymal cells does not occur to any significant degree[22, 36–38].

Using reassortant viruses containing distinct combinations of genes derived from T1 and T3, Weiner et al.[36, 37] showed that the major determinant of the pattern of CNS injury produced by reoviruses was the viral S1 gene, which encodes the outer capsid protein $\sigma 1$. An identical strategy has been used to show that reovirus tropism for retinal ganglion cells[40], for cells in the anterior lobe of the pituitary[41], and for isolated ependymal cells *in vitro*[42], is also determined by the S1 gene.

In order further to characterize the role of the $\sigma 1$ protein in reovirus CNS tropism, reoviruses with antigenically altered $\sigma 1$ proteins have been generated for both T3[43] and T1 (S. Lynn and B. N. Fields, unpublished). In both cases, the $\sigma 1$ variants were selected by their resistance to neutralizing anti-$\sigma 1$ monoclonal antibodies.

The T3 $\sigma 1$ variants have been extensively characterized[39,43-45]. All five variants characterized to date are at least 10 000-fold less virulent than the T3 strain from which they were derived[43,44]. In addition, whereas T3 generally achieves a peak brain titre following IC inoculation into newborn mice between 10^9 and 10^{11} plaque-forming units (PFU) per ml, the $\sigma 1$ variants achieve titres that are 3–5 $\log_{10}$ lower[43,45]. The brain pathology produced by the $\sigma 1$ variants also differs from that of T3. The variants produce extensive destruction and subsequent cavitation in the hippocampus and regions of the limbic system including the septum and mamillary bodies, however, unlike T3, they do not produce significant injury to the cortex[39,44].

The complete nucleotide sequence of the reovirus T3 (Dearing) S1 gene has been compared to that of five of the $\sigma 1$ antigenic variants. Each of the five variants has a single nucleotide change which would result in a single amino acid substitution in the $\sigma 1$ protein. In four of the five $\sigma 1$ variants the glutamic acid at position 419 is changed to either a lysine, alanine or glycine. In one $\sigma 1$ variant, glutamic acid at 419 is unaltered but an aspartic acid at position 340 is changed to a valine[44].

These studies indicate that:

(1) the $\sigma 1$ variants contain single amino acid substitutions in the protein against which the selecting monoclonal antibodies are directed;

(2) the variants have markedly attenuated neurovirulence and restricted CNS cell tropism.

These studies did not prove that the identified protein changes in $\sigma 1$ cause the altered biological properties of the variants. In order conclusively to establish this point, the biological properties of a reassortant virus containing the S1 gene derived from a T3 $\sigma 1$ antigenic variant were compared to those of an analogous reassortant with the wild type S1 gene. These studies clearly indicated that the attenuated neurovirulence, restricted CNS cell tropism, and altered CNS growth properties of the variant were due to the single amino acid substitution in the $\sigma 1$ protein[45].

It has recently been possible to isolate and begin to characterize a reovirus T3 receptor on neural and lymphoid cells. Both syngeneic monoclonal and polyclonal anti-idiotypic antibodies directed against an anti-$\sigma 1$ monoclonal antibody ('G5')[46] specifically bind to a T3 receptor on murine R1.1 thymoma and rat B104 neuroblastoma cells[47]. These antibodies can block viral binding to these cells[47], to transformed lymphocytes[48], and to dissociated cell cultures of cortical neurons[49]. Using the polyclonal anti-idiotypic antibodies it was calculated that R1.1 and B104 cells possessed 50 000–78 000 reovirus binding sites with an apparent affinity (K_d) of 10^{-9} mol L^{-1}. By comparison, earlier studies using GH$_4$C$_1$ pituitary cells had suggested that these cells contained about 4000 viral binding sites of high affinity[50] ($K_d = 10^{-11}$ mol L^{-1}).

The T3 receptor was isolated from solubilized membrane extracts of R1.1 cells using the polyclonal anti-idiotypic antibodies. Both T3 and antibodies bind to a monomeric glycoprotein of 67 kDa size in solubilized lysates of either R1.1 cell membranes or B104 neuroblastoma cell membranes[47]. It was subsequently shown, using a variety of criteria, that the T3 receptor on these cells is structurally similar to the mammalian β-adrenergic receptor. First, the masses and isoelectric points of both receptors are identical. Second, the trypsin digests of both purified receptors display indistinguishable digestion patterns. Third, anti-reovirus receptor antibodies immunoprecipitate purified β-adrenergic receptor. Fourth, both β-agonists and β-antagonists bind to purified reovirus T3 receptor[51].

The tropism of T3 in the CNS is determined in large part by the interaction of the viral σ 1 protein with a specific receptor present on neuronal cells. The resulting lethal encephalitis also insures that the S1 gene is a major determinant of T3 virulence[36, 37]. However, additional studies have clearly shown a role for host factors[52] and other viral genes[53] in reovirus neurovirulence. Following intracerebral inoculation into mice, T3 produces a lethal encephalitis up to 8–10 days of age. Mice at this age or older either develop non-lethal encephalitis or no clinical or histological signs of illness. The diminution of viral virulence in older animals correlates with diminished replication of virus in neural tissues[52]. The mechanism(s) responsible for the age-dependent susceptibility of mice to reovirus T3 infection remain unclear.

An appreciation of the role played by outer capsid proteins other than σ 1 in determining reovirus neurovirulence came from studies of natural isolates of T3. Different field isolates of T3 differ at least 1000-fold in their relative neurovirulence in newborn mice. Using reassortant viruses it has been shown that the viral M2 gene segment, which encodes the major outer capsid protein μ 1c, is the major determinant of relative avirulence within viruses of the T3 serotype. The mechanism by which the μ 1c protein affects neurovirulence remains unclear, although it does not appear to act by altering viral tropism within the CNS. The avirulent T3 isolates reach peak brain titres that are at least 100-fold less (e.g. 5×10^7 PFU per brain) than those of virulent isolates, suggesting that one factor in the diminished virulence may be decreased capacity to grow in neural tissue *in vivo*[53]. Thus, although the S1 gene is the major determinant of viral tropism within the CNS, other viral genes, such as M2, clearly play a significant role in determining relative neurovirulence within viruses belonging to the T3 serotype.

CONCLUSION

In this brief review we have attempted to summarize many of the major features of reovirus infections in the CNS. Particular emphasis has been placed on reviewing specific stages in the pathogenesis of reovirus infections of the nervous system, and in highlighting areas where studies with reoviruses have led to a better understanding of the pathogenesis of viral infections of the CNS in terms of individual viral genes and the proteins they encode.

Acknowledgements

Kenneth Tyler is the recipient of a physician-scientist award (AI00610) from the NIAID, and is an Alfred P. Sloan Research Fellow. Research support includes Program Project Grant 2 PO1 NS16998 from the NINCDS, Research Grant 5 RO1 AI13178 from the NIAID, and additional support from the Shipley Institute of Medicine.

References

1. Fields, B. N. and Greene, M. I. (1982). Genetic and molecular mechanisms of viral pathogenesis: implications for prevention and treatment. *Nature*, **300**, 19–23
2. Tyler, K. L. and Fields, B. N. (1987). Reovirus. In Murphy, F. A. (ed.) *The Laboratory Diagnosis of Infectious Diseases. Virology*. Chapter 21. (New York: Springer Verlag) (In press)
3. Sharpe, A. H. and Fields, B. N. (1983). Pathogenesis of reovirus infection. In Joklik, W. K. (ed.) *The Reoviridae*. pp. 229–85. (New York: Plenum Publishing Corporation)
4. Sharpe, A. H. and Fields, B. N. (1985). Pathogenesis of viral infections. Basic concepts derived from the reovirus model. *N. Engl. J. Med.*, **312**, 486–97
5. Tyler, K. L. and Fields, B. N. (1985). Reovirus and its replication. In Fields, B. N. (ed.) *Virology*. pp. 823–62 (New York: Raven Press)
6. Johnson, R. T. (1982). *Viral Infections of the Nervous System*. (New York: Raven Press)
7. Mims, C. A. and White, D. O. (1984). *Viral Pathogenesis and Immunology*. (Oxford: Blackwell Scientific Publications)
8. Wolf, J. L., Rubin, D. H., Finberg, R., Kauffman, R. S., Sharpe, A. H., Trier, J. S. and Fields, B. N. (1981). Intestinal M cells: A pathway for entry of reovirus into the host. *Science*, **212**, 471–2
9. Wolf, J. L., Kauffman, R. S., Finberg, R., Dambrauskas, R., Fields, B. N. and Trier, J. S. (1983). Determinants of reovirus interaction with the intestinal M cells and absorptive cells of murine intestine. *Gastroenterology*, **85**, 291–300
10. Kauffman, R. S., Wolf, J. L., Finberg, R., Trier, J. S. and Fields, B. N. (1983a). The σ 1 protein determines the extent of spread of reovirus from the gastrointestinal tract of mice. *Virology*, **124**, 403–10
11. Bodian, J. D. (1956). Poliovirus in chimpanzee tissues after virus feeding. *Am. J. Hyg.*, **64**, 181–97
12. Rubin, D. H. and Fields, B. N. (1980). Molecular basis of reovirus virulence. Role of the M2 gene. *J. Exp. Med.*, **152**, 853–68
13. Keroack, M. and Fields, B. N. (1986). Viral shedding and transmission between hosts determined by reovirus L2 gene. *Science*, **232**, 1635–8
14. Drayna, D. and Fields, B. N. (1982). Biochemical studies on the mechanism of chemical and physical inactivation of reovirus. *J. Gen. Virol.*, **63**, 161–70
15. Drayna, D. and Fields, B. N. (1982). Genetic studies on the mechanism of chemical and physical inactivation of reovirus. *J. Gen. Virol.*, **63**, 149–59
16. Sabin, A. B. and Olitsky, D. K. (1937, 1938). Influence of host factors on neuroinvasiveness of vesicular stomatitis virus. *J. Exp. Med.*, **66**, 15–34, 35–57; **67**, 201–28, 229–49
17. Sabin, A. B. (1956). Pathogenesis of poliomyelitis, reappraisal in light of new data. *Science*, **123**, 1151–7
18. Goodpasture, E. W. (1925). The pathways of infection of the central nervous system in herpetic encephalitis of rabbits contracted by contact; with a comparative comment on medullary lesions in a case of human poliomyelitis. *Am. J. Pathol.*, **1**, 29–46
19. Tyler, K. L., McPhee, D. A. and Fields, B. N. (1986). Distinct pathways of viral spread in the host determined by reovirus S1 gene sement. *Science*, **233**, 770–4
20. Papadimitriou, J. M. (1967). An electron microscopic study of reovirus 3 encephalitis. *Am. J. Pathol.*, **50**, 59–75

21. Gonatas, N. K., Margolis, G. and Kilham, L. (1971). Reovirus type III encephalitis: observations of virus-cell interactions in neural tissues. II. Electron microscopic studies. *Lab. Invest.*, **24**, 101–9
22. Raine, C. S. and Fields, B. N. (1973). Reovirus type III encephalitis – A virologic and ultrastructural study. *J. Neuropathol. Exp. Neurol.*, **32**, 19–33
23. Hassan, S. A., Rabin, E. R., and Melnick, J. L. (1965). Reovirus myocarditis in mice: An electron microscopic, immunofluorescent and virus assay study. *Exp. Mol. Pathol.*, **4**, 66–80
24. Jenson, A. B., Rabin, E. R., Bentinck, D. C. and Rapp, F. (1966). Reovirus viremia in newborn mice. *Am. J. Pathol.*, **49**, 1171–83
25. Kundin, W. D., Liu, C. and Gigstad, J. (1966). Reovirus infection in suckling mice: immunofluorescent and infectivity studies. *J. Immunol.*, **97**, 393–401
26. Oakes, J. E., Gray, W. L. and Lausch, R. N. (1986). Herpes simplex virus type 1 DNA sequences which direct spread of virus from cornea to central nervous system. *Virology*, **150**, 513–17
27. Thompson, R. L., Cook, M. L., Devi-Rao, G. B., Wagner, E. K. and Stevens, J. G. (1986). Functional and molecular analyses of the avirulent wild-type herpes simplex virus type 1 strain KOS. *J. Virol.*, **58**, 203–11
28. Margolis, G. and Kilham, L. (1969). Hydrocephalus in hamsters, ferrets, rats, and mice following inoculations with reovirus type I. II. Pathologic studies. *Lab. Invest.*, **21**, 189–98
29. Wolinsky, J. S., Klassen, T. and Baringer, J. R. (1976). Persistence of neuroadapted mumps virus in brains of newborn hamsters after intraperitoneal inoculation. *J. Infect. Dis.*, **133**, 260–7
30. Johnson, R. T. (1965). Virus invasion of the central nervous system. A study of Sindbis virus infection in the mouse using fluorescent antibody. *Am. J. Pathol.*, **46**, 929–43
31. Weiner, H. L., Ault, K. A. and Fields, B. N. (1980). Interaction of reovirus with cell surface receptors. I. Murine and human lymphocytes have a receptor for the hemagglutinin of reovirus type 3. *J. Immunol.*, **124**, 2143–8
32. Haase, A. T. (1986). Pathogenesis of lentivirus infections. *Nature*, **322**, 130–6
33. Summers, B. A., Griesen, H. A. and Appel, M. J. (1978). Possible initiation of viral encephalitis in dogs by migrating lymphocytes infected with distemper. *Lancet*, **1**, 187–9
34. Nielsen, S. L. and Baringer, J. R. (1972). Reovirus-induced aqueductal stenosis in hamsters. *Lab. Invest.*, **27**, 531–7
35. Masters, C., Alpers, M. and Kakulas, B. (1977). Pathogenesis of reovirus type 1 hydrocephalus in mice. *Arch. Neurol.*, **34**, 18–28
36. Weiner, H. L., Powers, M. L. and Fields, B. N. (1980). Absolute linkage of virulence and central nervous system cell tropism of reoviruses to viral hemagglutinin. *J. Infect. Dis.*, **141**, 609–16
37. Weiner, H. L., Drayna, D., Averill, D. R., Jr. and Fields, B. N. (1977). Molecular basis of reovirus virulence: Role of the S1 gene. *Proc. Natl. Acad. Sci. USA*, **74**, 5744–8
38. Margolis, G., Kilham, L. and Gonatas, N. K. (1971). Reovirus type III encephalitis: observations of virus-cell interactions in neural tissues. I. Light microscopy studies. *Lab. Invest.*, **24**, 91–100
39. Spriggs, D. R., Bronson, R. T. and Fields, B. N. (1983). Haemagglutinin variants of reovirus type 3 have altered central nervous system tropism. *Science*, **220**, 505–7
40. Tyler, K. L., Bronson, R. T., Byers, K. B. and Fields, B. N. (1985). Molecular basis of viral neurotropism: Experimental reovirus infection. *Neurology*, **35**, 88–92
41. Onodera, T., Toniolo, T., Ray, U. R., Jensen, A. B., Knazek, R. A. and Notkins, A. L. (1981). Virus-induced diabetes mellitus. XX. Polyendocrinopathy and autoimmunity. *J. Exp. Med.*, **153**, 1457–73
42. Tardieu, M. and Weiner, H. L. (1982). Viral receptors on isolated murine and ependymal cells. *Science*, **215**, 419–21
43. Spriggs, D. R. and Fields, B. N. (1982). Attenuated reovirus type 3 strains generated by selection of haemagglutinin antigenic variants. *Nature*, **297**, 68–70
44. Bassel-Duby, R., Spriggs, D. R., Tyler, K. L. and Fields, B. N. (1986). Identification of attenuating mutations on the reovirus type 3 S1 double-stranded RNA segment with a rapid sequencing technique. *J. Virol.*, **60**, 64–7
45. Kaye, K. M., Spriggs, D. R., Bassel-Duby, R., Fields, B. N. and Tyler, K. L. (1986). Genetic

18

basis for altered pathogenesis of an immune-selected antigenic variant of reovirus type 3 (Dearing). *J. Virol.*, **59**, 90–7

46. Nepom, J. T., Weiner, H. L., Dichter, M. A., Tardieu, M., Spriggs, D. R., Gramm, C. F., Powers, M. L., Fields, B. N. and Greene, M. I. (1982). Identification of a hemagglutinin-specific idiotype associated with reovirus recognition shared by lymphoid and neural cells. *J. Exp. Med.*, **155**, 155–67

47. Co, M. S., Gaulton, G. N., Fields, B. N. and Greene, M. I. (1985). Isolation and biochemical characterization of the mammalian reovirus type 3 cell-surface receptor. *Proc. Natl. Acad. Sci. USA*, **82**, 1494–8

48. Kauffman, R. S., Noseworthy, J. H., Nepom, J. T., Finberg, R., Fields, B. N. and Greene, M. I. (1983). Cell receptors for the mammalian reovirus. II. Monoclonal anti-idiotypic antibody blocks viral binding to cells. *J. Immunol.*, **131**, 2539–41

49. Dichter, M. A., Weiner, H. L., Fields, B. N., Mitchell, G., Noseworthy, J., Gaulton, G. and Greene, M. (1986). Anti-idiotypic antibody to reovirus binds to neurons and protects from viral infection. *Ann. Neurol.*, **19**, 555–8

50. Maratos-Flier, T., Kahn, C. R., Spriggs, D. R. and Fields, B. N. (1983). Specific plasma membrane receptors for reovirus on rat pituitary cells in culture. *J. Clin. Invest.*, **72**, 617–21

51. Co, M. S., Gaulton, G. N., Tominaga, A., Homcy, C. J., Fields, B. N. and Greene, M. I. (1985). Structural similarities between the mammalian β-adrenergic and reovirus type 3 receptors. *Proc. Natl. Acad. Sci. USA*, **82**, 5315–18

52. Tardieu, M., Powers, M. L. and Weiner, H. L. (1983). Age dependent susceptibility to reovirus type 3 encephalitis: Role of viral and host factors. *Ann. Neurol.*, **13**, 602–7

53. Hrdy, D. B., Rubin, D. H. and Fields, B. N. (1982). Molecular basis of reovirus virulence: Role of the M2 gene in avirulence. *Proc. Natl. Acad. Sci. USA*, **79**, 1298–302

3
Dysmyelination in Transgenic Mice Containing the Early Region of JC Virus

B. D. TRAPP, J. A. SMALL AND G. A. SCANGOS

INTRODUCTION

JC virus is a human papovavirus that has been implicated as the causative agent in the chronic human demyelinating disease progressive multifocal leukoencephalopathy (PML). Although the virus is widespread in human populations[1,2], PML in a rare opportunistic disease that occurs primarily in patients with compromised immune systems[3]. The incidence of PML has increased recently in association with HIV infection[4]. Numerous studies have shown viral particles within abnormal oligodendrocytes from PML brains[5] and the aetiology of demyelination is thought to be due to lytic destruction of these myelin-forming cells. Astrocytes can have bizarre morphologies in PML brains that include some features of transformed cells. Virions are rarely found in electron micrographs of these morphologically altered cells[5].

When inoculated into newborn hamsters JC virus causes a variety of tumours[6–8] derived primarily from neural tissue. JC virus is the only human virus known to cause tumours in primates (owl and squirrel monkeys) where grade IV astrocytomas or glioblastomas develop 18–24 months after intracranial injection[9]. JC virus has not been shown to induce any detectable pathology when injected into mice[10]. Therefore experimental models addressing the possible mechanism of JC virus-induced demyelination have not been available.

To bypass host range restrictions and to obtain JC virus sequences in every cell of the mouse, transgenic mice have been produced by introducing the early region of the JC virus genome into fertilized mouse ova[11]. The purpose of the present chapter is to describe morphological alterations that occur in strains of JCV transgenic mice. In addition, the molecular basis for demyelination in the most severely affected strain (JC-48) of JCV transgenic mice is reviewed and the role that JC virus T-antigen expression may have in the pathogenesis of dysmyelination is discussed.

TRANSGENIC MICE

JC virus early region DNA was micro-injected into pronuclei of fertilized mouse eggs as described by Gordon and Ruddle[12]. The early region of JC virus contains the origin of replication, the transcriptional control region and the genes encoding large and small T-antigens. Structural viral proteins are encoded in the late region and were not included on the constructs injected. Therefore, no productive viral infection can occur. T-antigens, the only JC viral proteins that can be expressed in these mice, are pleiotropic transforming proteins found in both the nucleus and cytoplasm of papovavirus transformed cells[10]. T-antigens appear to play a role in regulating DNA replication and gene expression and are also thought to have a number of functions within the cytoplasm[10].

Ten animals containing complete copies of JCV early region DNA were identified[11, 13]. Five of these mice were born dead or died shortly after birth. No analyses were performed on these mice. The other five animals survived varying lengths of time and expressed JCV T-antigens. Three females developed adrenal medullary neuroblastomas and died at approximately 14 weeks of age. Two males survived up to 10 months of age and are the basis of the lines established. One line (strain 48) is derived from a mosaic founder mouse (incorporation of JCV DNA occurred after the pronuclei stage) in which 20% of the offspring contain intact JC virus early region DNA in all cells. The other line (strain 91) is derived from a non-mosaic JC founder in which 50% of the offspring contain intact JC virus early region DNA in all cells.

BEHAVIOURAL PHENOTYPE

The offspring of JC-48 appeared normal at birth. However, at 2 weeks of age approximately 20% of the animals developed a severe tremor that was readily apparent when the mice were moving, but not when they were at rest. Analysis of tail DNA for JC virus early region by the Southern blot technique revealed that only the mice which displayed the neurological phenotype contained intact copies of JC virus early region[11]. The affected JC-48 mice began to exhibit tonic seizures at 3 weeks of age, most died by 4 weeks and none lived past 6 weeks of age. Since no affected mice lived to maturity, a permanent line of JC-48 mice was not established.

Approximately 50% of the offspring of JC-91 exhibited a similar but less severe neurological phenotype. A slight tremor became evident 3–4 weeks after birth. Only those JC-91 mice displaying the neurological phenotype contained intact copies of JC virus early region. Affected JC-91 mice were healthier than JC-48 mice and lived to 2–5 months of age. JC-91 mice have been successfully bred and homozygous offspring identified. The severity of the neurological dysfunction in JC-91 homozygotes is intermediate to that found in JC-48 and JC-91 heterozygotes. JC-91 homozygotes express a shaking phenotype at 2 weeks of age and none live past 4 months of age.

DEGREE OF T-ANTIGEN EXPRESSION AND DYSMYELINATION ARE RELATED

All available evidence indicates that the dysmyelination in JCV transgenic mice is related to the expression of JC virus T-antigens. The correlation between dysmyelination and inheritance of intact copies of JC virus early region is absolute. Integration of JC virus DNA into the genome of JC-48 and JC-91 mice occurred at different sites[11]. The fact that they had similar dysmyelinating phenotypes indicates that expression of JC virus early region rather than insertional mutagenesis is reponsible for the dysmyelination. The level of T-antigen mRNA expression is greater in JC-48 brain than in JC-91 brain (Figure 3.1). The level of T-antigen mRNA expression in JC-91 homozygotes is intermediate to that found in JC-48 and JC-91 heterozygous mice. Therefore, the level of T-antigen expression in transgenic mice correlates with the severity of neurological dysfunction.

The level of JCV T-antigen mRNA expression is also related to the severity of dysmyelination in JCV transgenic mice (Figure 3.2). The number and diameter of myelinated fibres found in light micrographs of JC-91 heterozygous mice (Figure 3.2B) are similar to those found in controls (Figure 3.2A). Morphological alterations in these myelinated fibres were not always apparent in light micrographs. In contrast, the CNS of JC-91 homozygotes is hypomyelinated (Figure 3.2C). Myelin sheaths surround most of the larger axons; many of the smaller axons appear unmyelinated. Myelin sheaths are rarely found in light micrographs from JC-48 mice (Figure 3.2D). The

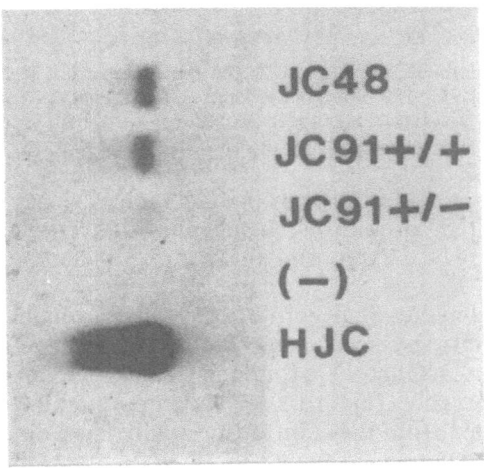

Figure 3.1 T-antigen RNA expression in transgenic mice. Total brain RNA was isolated from 4-week-old mice and polyadenylated RNA was selected by oligo(dT)-cellulose chromatography. 3 µg of poly (A+) RNA was electrophoresed in a formaldehyde/agarose gel, transferred to nitrocellulose, and hybridized with a JCV early region probe. JC-48 mice contain the highest levels of T-antigen mRNA (lane 1), JC-91 homozygotes intermediate levels (lane 2) and JC-91 heterozygotes the lowest levels (lane 3). Control mice do not contain T-antigen mRNA (lane 4). Cells from a hamster neuroblastoma that were infected with JC virus served as a positive control (lane 5). (With permission from the Editorial Offices of *Cell*)

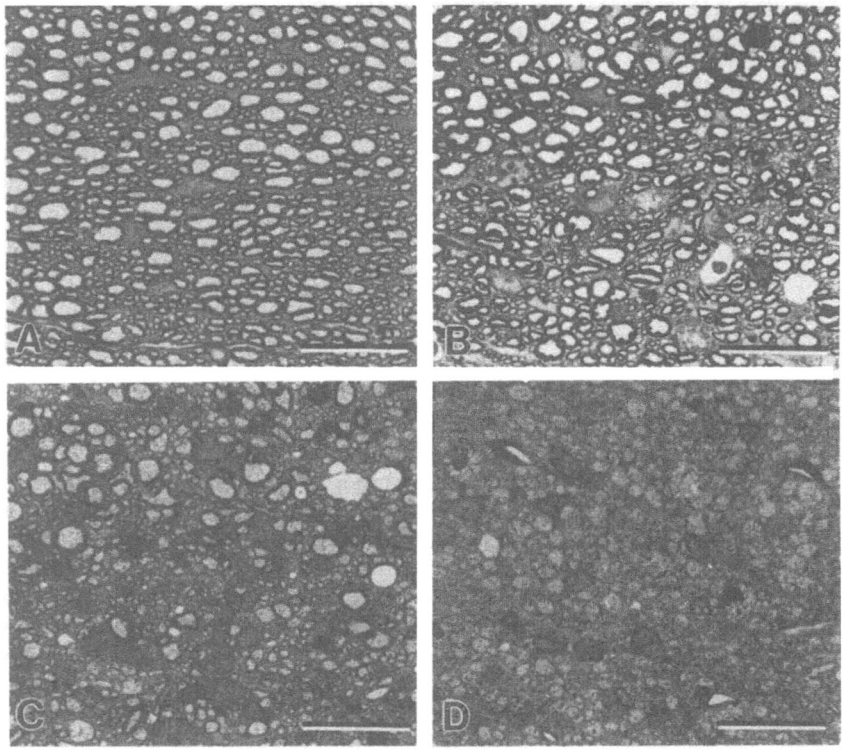

Figure 3.2 1 µm thick Epon sections of the spinal cord from 28-day-old control (A), JC-91 heterozygous (B), JC-91 homozygous (C) and JC-48 (D) mice. The severity of dysmyelination in JC transgenic mice is related to the level of JCV T-antigen mRNA expression (see Figure 3.1) and the severity of neurological dysfunction. Scale bars = 25 µm

number and diameter of axons in JC-48 mice are not dramatically reduced when compared to controls. In all lines of transgenic mice, neurons do not appear to be altered and the peripheral nervous system is normally myelinated.

Sections from the spinal cord of JC-91 heterozygous and JC-48 transgenic mice were studied by electron microscopy. Vesicular disruption and vacuolation of myelin lamellae were found in electron micrographs of sections from JC-91 heterozygotes (Figure 3.3). The inner margins of myelin sheaths are often separated from the axon by unusually large distances. These findings indicated that oligodendrocytes in JC-91 heterozygous mice formed thick myelin sheaths but were unable properly to maintain them. In electron micrographs of sections from JC-48 mice, most axons are surrounded by cellular processes (Figure 3.4). Some of these processes encircle the axon several times; their membranes are closely opposed and they contain profiles similar to outer and inner tongue processes and paranodal specialization of myelin sheaths. Some of these processes extend from cells that ultrastructurally resembled immature oligodendrocytes. These observations indicate that

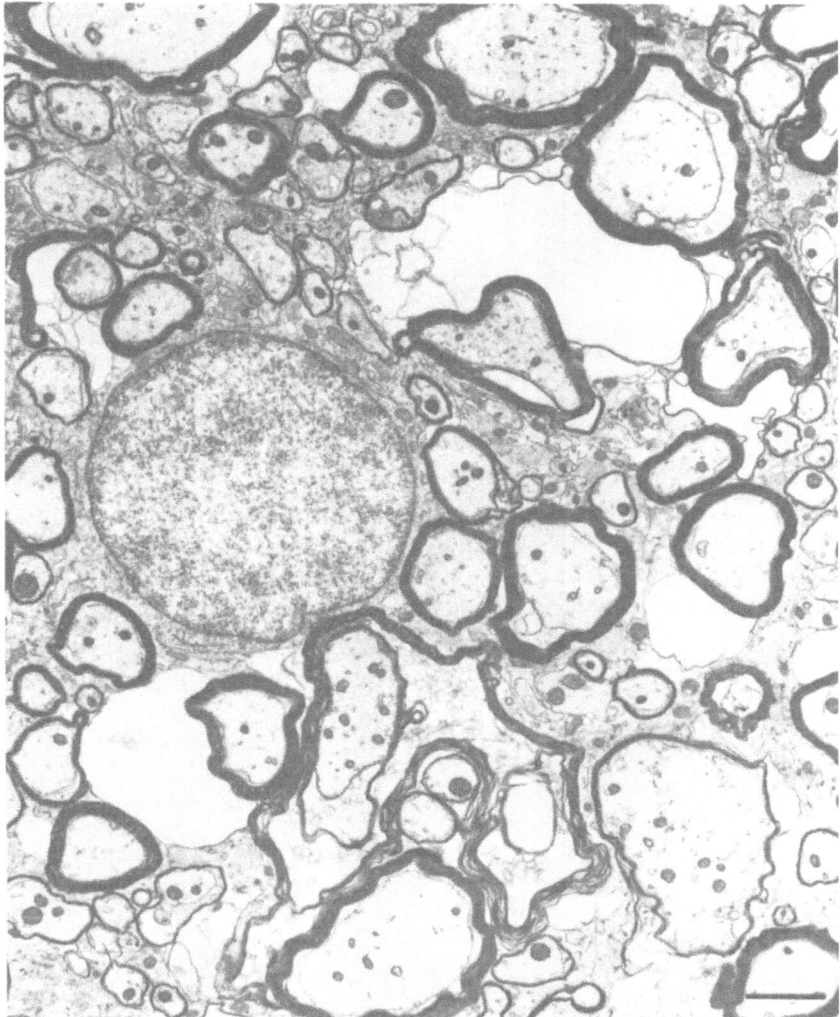

Figure 3.3 Electron micrograph of an oligodendrocyte and myelinated axons from 28-day-old JC-91 heterozygous mouse. Vesiculation and vacuolation of myelin lamellae are present. Scale bar = 2 µm

JC-48 oligodendrocytes extend processes that ensheath, but fail to myelinate axons properly.

DYSMYELINATION IN JC-48 TRANSGENIC MICE

Once it was evident that JCV T-antigen expression was the likely cause of dysmyelination in these transgenic mice we began to study how these molecules might affect oligodendrocyte function. It was essential to establish

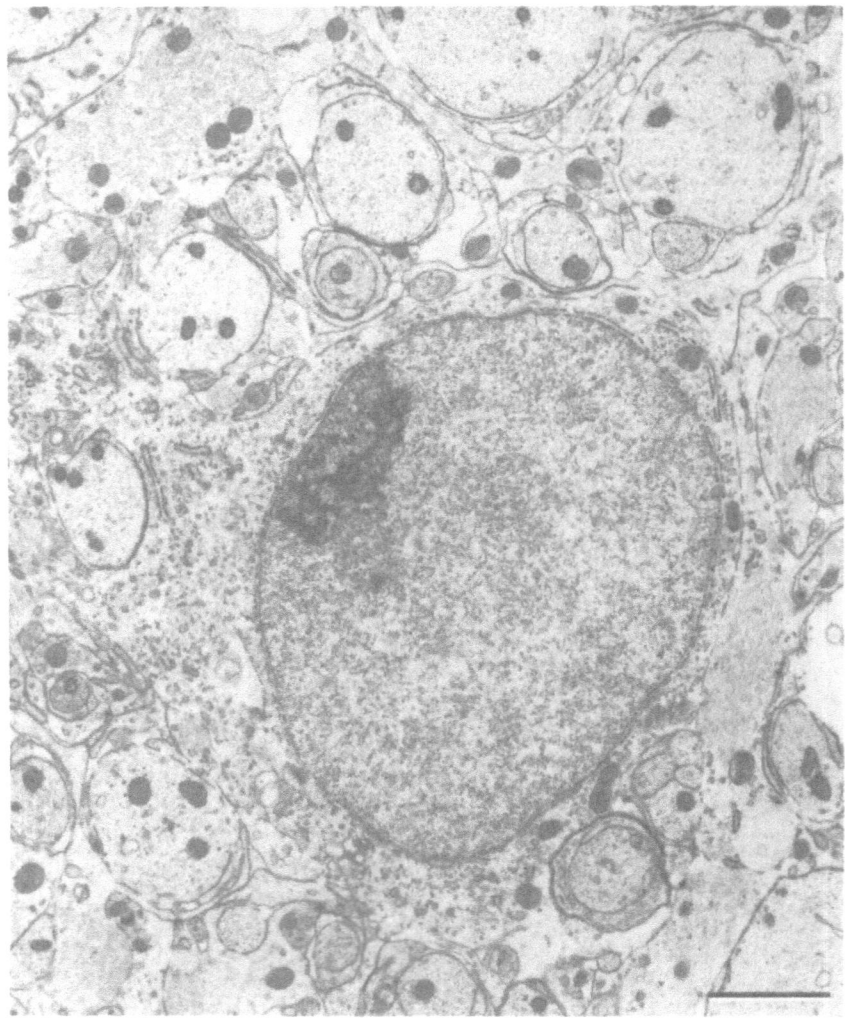

Figure 3.4 Electron mirograph of an oligodendrocyte and axons from a 28-day-old JC-48 mouse. Most axons are ensheathed by cellular processes. The membranes of some of these processes are tightly compacted and resemble thin myelin sheaths. The oligodendrocyte is immature in appearance. Scale bar = 10 µm

whether JC T-antigens were expressed in all brain cells or exclusively in oligodendrocytes. In addition, we were interested in whether myelin gene expression was altered in these animals. Using standard immunocyto-chemical[14] and *in situ* hybridization[15] procedures the distributions of JCV T-antigens mRNA and myelin protein gene products were determined in paraffin sections from 28-day-old JC-48 mice[16]. These mice were best suited for these studies because they are the most severely dysmyelinated.

Expression of JCV T-antigens

To determine where the early region of JC virus was expressed, sections of JC-48 brain were hybridized with JC virus early region DNA. This probe binds specifically to mRNA encoding JCV T-antigens[11]. In sections of JC-48 brain, JCV T-antigen mRNA was distributed over the same regions as proteolipid (PLP) and myelin basic protein (MBP) mRNAs (Figure 3.5).

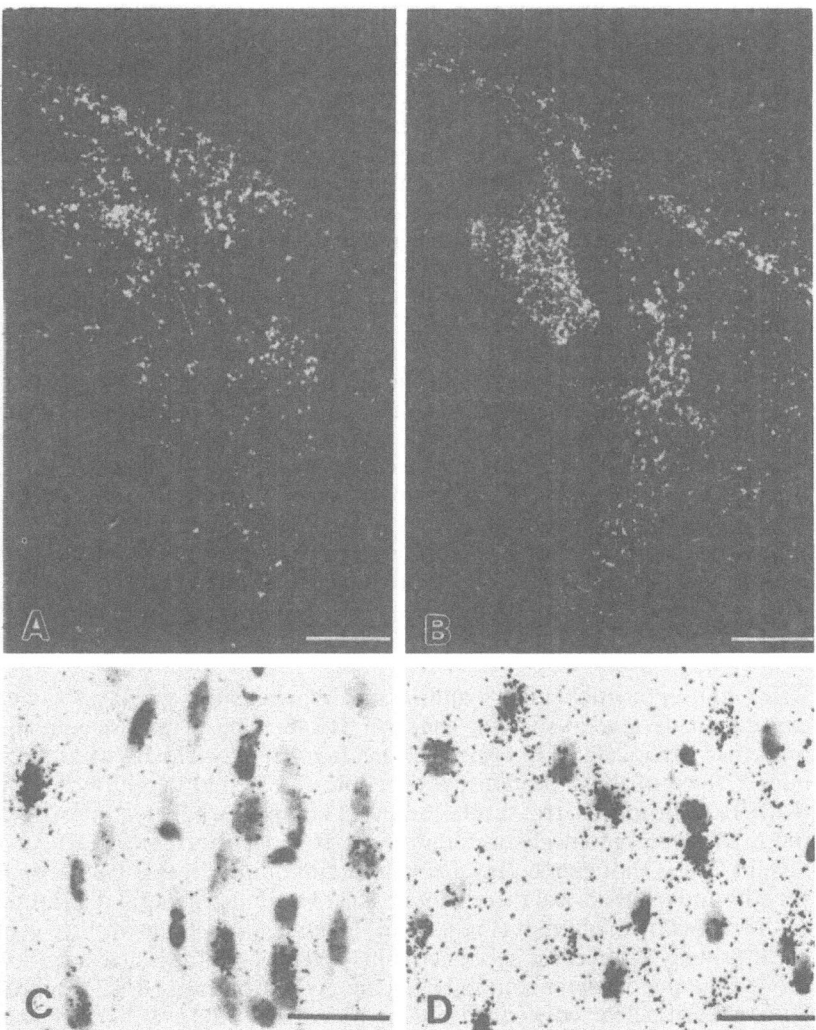

Figure 3.5 The distribution of JCV T-antigen mRNA in sections of the telencephalon from 28-day-old JC-48 mice (A, C) is similar to the distribution of PLP mRNA in sections from age-matched controls (B, D). Both mRNAs appear in clusters around oligodendrocyte nuclei. Silver grains produced by hybridizing probes in A and B are visualized as bright spots under dark field optics. C and D bright field images. Scale bars: A, B = 350 μm; C, D = 25 μm

JCV T-antigen mRNAs are concentrated within cell perikarya that have the distribution and appearance of oligodendrocytes in both white and grey matter. JC T-antigen mRNA was not detected in neurons nor in astrocytes of the glia limitans.

These studies established that JCV T-antigens were expressed predominantly in JC-48 oligodendrocytes. Therefore, T-antigen expression by oligodendrocytes is the likely cause of dysmyelination in JC-48 mice.

Viral life cycles are regulated at different stages that include: viral absorption, penetration, uncoating, transcription, translation and assembly. Since JC-48 transgenic mice received the JCV sequence through the germ line, they contained the viral sequence in every cell thereby circumventing the early stages of the virus life cycle from absorption to uncoating. In addition, the virus life cycle could not go to completion because the DNA constructs did not contain the late region of the viral genome. Therefore, specific expression of the JCV T-antigens in oligodendrocytes must be mediated at the level of gene expression. Thus, the transgenic mice have bypassed early levels of restriction of viral infection and demonstrate that regulation of gene expression alone is sufficient to direct virally-induced pathology predominantly in oligodendrocytes.

The mechanisms of viral gene regulations are likely to be complex. Previous work demonstrated that the JC viral enhancer can elevate the level of expression of linked genes in primary human fetal glial cells[17], suggesting that JCV expression can be regulated by positive factors. Recent data (A. Beggs and G. Scangos, unpublished) have demonstrated the presence of factors that can extinguish expression of JCV in many fibroblast cell lines, suggesting that negative regulation is also operative. Thus, it seems likely that expression of JCV predominantly in oligodendrocytes is mediated through both positive and negative genetic regulatory mechanisms.

Myelin gene expression

Using immunocytochemistry and *in situ* hybridization we have studied the expression of myelin-specific genes in JC-48 mice. Three myelin-specific proteins were localized in brain sections from 28-day-old JC-48 and control mice[16]. Two of these proteins, proteolipid (PLP) and myelin basic protein (MBP) are the major structural proteins of compact CNS myelin[18]; the other, the myelin-associated glycoprotein (MAG) is a minor constituent of CNS myelin that is thought to play a role in maintaining contact between myelin sheaths and axons[19]. All three proteins have a similar general distribution in JC-48 and control animals (Figure 3.6). However, the levels of these proteins are markedly reduced in sections from JC-48 mice. Perinuclear regions of JC-48 oligodendrocytes contained high concentrations of MAG (Figure 3.7). Similar perinuclear concentrations of MAG are found in control sections from 5- and 10-day-old mice, but not in sections from 28-day-old controls[16].

To investigate the molecular basis for the reduction in myelin protein expression in JC-48 mice, the levels and distributions of PLP and MBP mRNAs were studied in paraffin sections by *in situ* hybridization using ^{35}S-labelled cDNA probes. The distribution and intensity of PLP and MBP

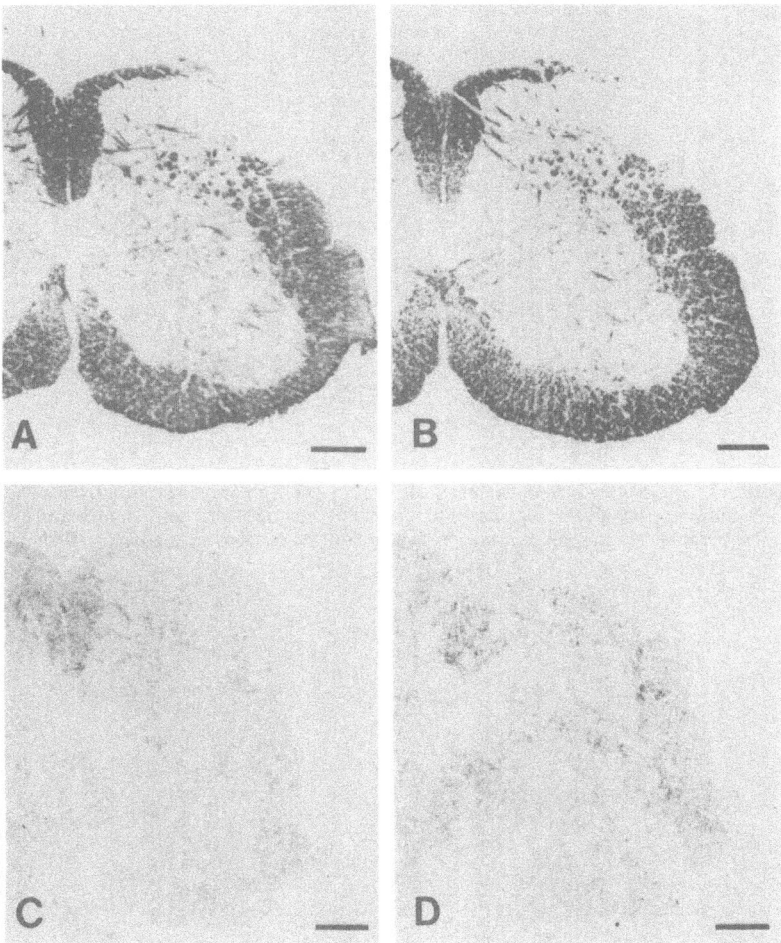

Figure 3.6 Comparison of the immunocytochemical localization of MBP (A, C) and PLP (B, D) in paraffin sections from 28-day-old control (A, B) and JC-48 (C, D) spinal cord. MBP and PLP are similarly distributed but dramatically reduced in JC-48 mice. Scale bars = 10 μm

hybridization signal was similar in JC-48 and control sections (Figures 3.8 and 3.9). PLP mRNA is concentrated around oligodendrocyte nuclei whereas MBP mRNA is distributed diffusely over white matter tracts.

These results establish that myelin-specific genes are expressed throughout the CNS of JC-48 mice. However, there is a dramatic discordance between the levels of transcriptional and translational products of the MBP and PLP genes (Figures 3.8 and 3.9). These observations have been confirmed by Northern and Western blots (J. Small and G. Scangos, unpublished observations). Therefore dysmyelination in JC-48 mice is not due to the failure of oligodendrocytes to activate myelin-specific genes. Northern blot analyses of

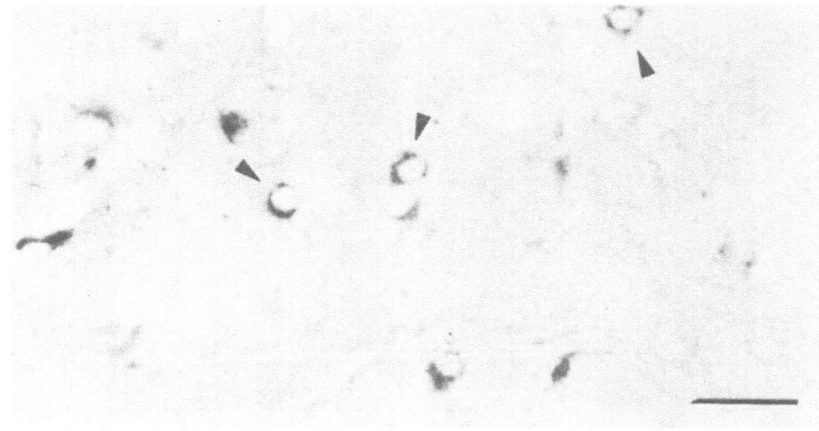

Figure 3.7 Immunocytochemical localization of MAG in a paraffin section from a 28-day-old JC-48 mouse. Perinuclear regions of oligodendrocytes are intensely stained (arrowheads). Similar perinuclear staining was not present in control sections. Scale bar = 10 µm

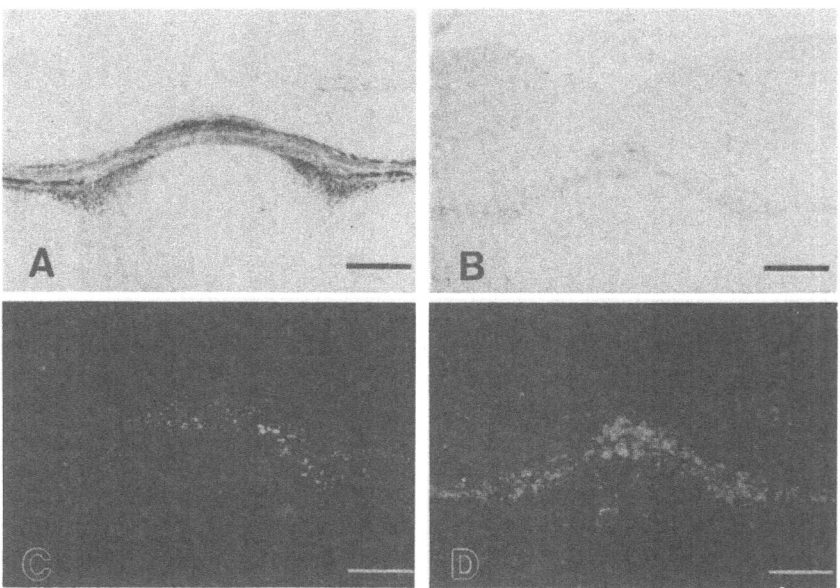

Figure 3.8 Comparison of the distributions of PLP (A, B) and PLP mRNA (C, D) in paraffin sections of the corpus callosum from 28-day-old control (A, C) and JC-48 transgenic (B, D) mice. PLP is reduced in JC-48. PLP mRNA visualized as bright spots under dark field optics, is concentrated around oligodendrocyte nuclei and is detected at similar concentrations in control and JC-48 sections. Scale bars = 350 µm

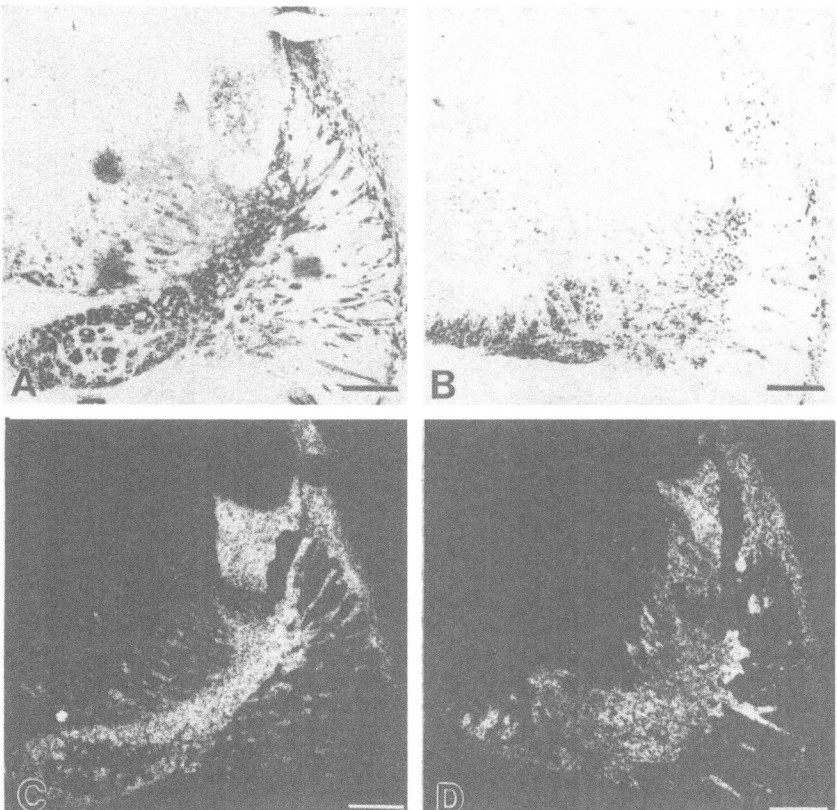

Figure 3.9 Comparison of the distributions of MBP (A, B) and MBP mRNA (C, D) in paraffin sections of the telencephalon from control (A, C) and JC-48 (B, D) mice. MBP is reduced in JC-48 mice. MBP mRNA is diffusely distributed over myelinated fibre tracts and detected at similar concentrations in control and JC-48 sections. Scale bars = 350 μm

RNA have shown PLP and MBP transcripts to be spliced to sizes with mobilities indistinguishable from those found in control animals. Possible explanations for the decreased ratio between transcriptional and translational products include decreased turnover of mRNA, decreased mRNA translation, increased turnover of protein and improper processing of protein.

The results described above indicated that the maturation of JC-48 oligodendrocytes was arrested at an early stage of development that permitted transcription of myelin genes, but not appropriate assembly of myelin. How JC virus T-antigen may be related to these changes is discussed below.

Astrocytes in JC-48 mice

Alterations in astrocytes within PML lesions are well documented and briefly discussed in the introduction of this chapter. Glial fibrillary acid protein

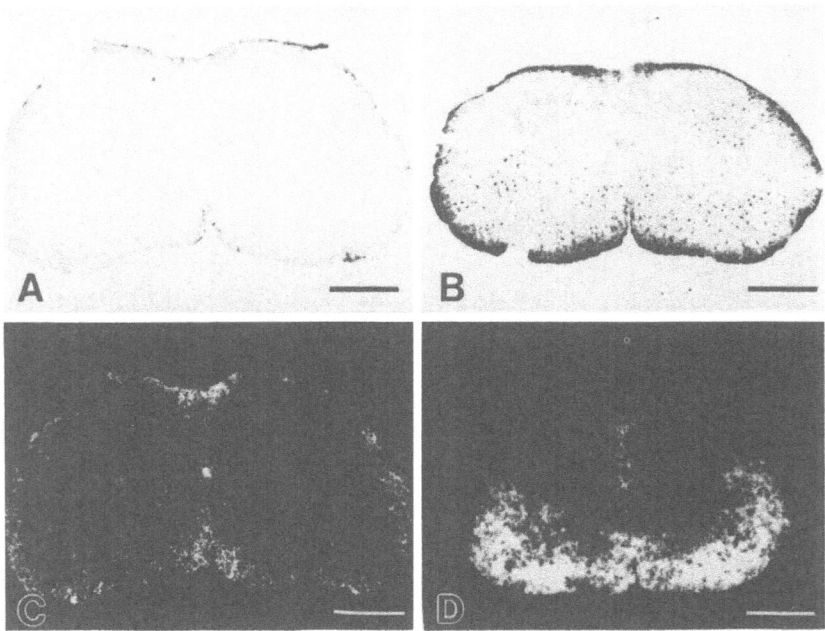

Figure 3.10 Comparison of the distribution of GFAP (A, B) and GFAP mRNA (C, D) in paraffin sections of spinal cord from control (A, C) and JC-48 (B, D) mice. Both GFAP and GFAP mRNA are increased in JC-48 mice. Scale bars = 350 μm

(GFAP) is a cytoskeletal protein that polymerizes to form the intermediate filaments of astrocytes[20]. Immunolocalization of GFAP demonstrates a hypertrophy of astrocytes in JC-48 transgenic mice (Figure 3.10). Astrocytic processes laden with GFAP intermediate filaments occupy much of the space normally filled by myelin. GFAP mRNA is also increased in JC-48 mice. The astrocytic response in JC-48 mice resembles that found in similarly hypomyelinated animals[21]. Our studies described above have indicated that astrocytes of the glia limitans and grey matter do not express detectable amounts of T-antigen mRNA. Therefore, we consider it unlikely that JCV T-antigens transform astrocytes in JC-48 mice and consider the hypertrophy of astrocytes to be a general response to the hypomyelinated environment.

ROLE OF T-ANTIGENS IN DYSMYELINATION

Our results suggest that JC virus T-antigens arrest the maturation of JC-48 oligodendrocytes at an early stage of development. Although transgenic oligodendrocytes do not properly myelinate axons, they expressed myelin-specific genes and display some of the morphological phenotypes associated with early stages of myelination.

Oligodendrocytes in JC-48 mice were identified by their morphological appearance in light and electron micrographs and by the perinuclear labelling of white matter cells by PLP cDNA and MAG antiserum. The number of PLP mRNA-positive cells per unit area of JC-48 white matter is not dramatically reduced compared to controls. Therefore, a paucity of oligo-dendrocytes does not appear to be responsible for the dysmyelination. JC-48 oligodendrocytes extended processes that ensheath axons. The diffuse distribution of PLP, MBP and MAG and of MBP mRNA within white matter tracts indicated that this ensheathment was extensive and that oligodendrocytes had the ability to transport gene products out to their processes. These processes can form thin spiral wraps that ultrastructurally resemble myelin.

Keeping in perspective what is known about papovavirus T-antigens in general, we can speculate about how T-antigens may affect the behaviour of oligodendrocytes. The T-antigens of papovaviruses are pleiotropic proteins that seem to have several different activities. Large T-antigen is both necessary and sufficient for transformation of primary established cell lines[10]. It also is required for viral DNA replication and for the regulation of both viral and cellular genes[10, 22, 23]. T-antigen binds to specific sites on DNA, and can form complexes with the cellular protein p53[23]. T-antigen has ATPase activity, which is necessary for DNA replication[24, 25]. Evidence has also been presented that the T-antigen of SV40 possesses DNA helicase activity[26].

The pleiotropic transforming activity of T-antigens suggests that expression in oligodendrocytes may interfere with differentiation and lead to proliferation by altering the cellular programme of gene expression. In this hypothesis, T-antigen may not specifically interfere with myelination but may alter the phenotype of oligodendrocytes by blocking or retarding their differentiation. Alternatively, since T-antigens bind to DNA and can regulate gene expression, it is possible that they interfere specifically with the expression of genes necessary for myelination. Since MBP and PLP mRNAs were present at approximately control levels in affected oligodendrocytes, it is unlikely that myelination was affected by interference with transcription of these genes. It is possible that expression of other genes necessary for myelination was inhibited and that feedback mechanisms exist that retarded translation of MPB and PLP mRNAs.

CONCLUDING REMARKS

Oligodendrocyte-specific expression of JC viral constructs in transgenic mice has provided an experimental model for investigating the role of JC T-antigens in dysmyelination. Since JCV T-antigen expression in mice can interfere with oligodendrocyte maturation and myelination, JCV infectivity of oligodendrocytes in patients with PML could potentially contribute to the demyelinating pathology by mechanisms other than lysis of oligodendrocytes. The introduction of genetic constructs into the germ line of mice should prove invaluable in elucidating the function of individual molecules during normal brain development and in CNS disease.

Acknowledgements

Bruce D. Trapp is a Harry Weaver Neuroscience Scholar of the National Multiple Sclerosis Society of America. George A. Scangos is a Leukemia Society of America Scholar. Judy A. Small is supported by a postdoctoral fellowship from the Cancer Research Institute, New York.

References

1. Gardner, S. D. (1973). Prevalence in England of antibody to human polyomavirus (BK). *Br. Med. J.*, **1**, 77-8
2. Padgett, B. L. and Walker, D. L. (1973). Prevalence of antibodies in human sera against JC virus, an isolate from a case of progressive multifocal leukoencephalopathy. *J. Infect. Dis.*, **127**, 467-70
3. Johnson, R. T. (1983). Evidence for polyomaviruses in human neurological diseases. In Sever, J. L. and Madden, D. L. (eds.) *Polyomaviruses and Human Neurological Disease*. pp. 183-90. (New York: Alan R. Liss)
4. Levy, R. M., Bredesen, D. E. and Rosenblum, M. L. (1985). Neurological manifestations of the acquired immunodeficiency syndrome (AIDS): Experience at UCSF and review of the literature. *J. Neurosurg.*, **62**, 475-95
5. ZuRhein, G. M. (1972). Virions in progressive multifocal leukoencephalopathy. In Minkler, J. (ed.) *Pathology of the Nervous System*. Vol. 3, pp. 2893-912. (New York: McGraw-Hill)
6. Walker, D. L., Padgett, B. L., ZuRhein, G. M., Albert, A. E. and Marsh, R. F. (1973). Human papovirus (JC): induction of brain tumors in hamsters. *Science*, **181**, 674-6
7. Padgett, B. L., Walker, D. L., ZuRhein, G. M. and Verakis, J. N. (1977). Differential neuro-oncogenicity of strains of JC virus, a human polyoma virus, in newborn Syrian hamsters. *Cancer Res.*, **37**, 718-20
8. Varakis, J., ZuRhein, G. M., Padgett, B. L. and Walker, D. L. (1978). Induction of peripheral neuroblastomas in Syrian hamsters after injection as neonates with JC virus, human polyoma virus. *Cancer Res.*, **38**, 1718-22
9. London, W. T., Houff, S. A., Madden, D. L., Fuccillo, D. A., Gravell, M., Wallen, W. C., Palmer, A. E., Sever, J. L., Padgett, B. L., Walker, D. L., ZuRhein, G. M. and Ohashi, T. (1978). Brain tumors in owl monkeys inoculated with a human polyoma virus (JC virus). *Science*, **102**, 1246-9
10. Tooze, J. (1981). *The Molecular Biology of Tumor Viruses*. (Cold Spring Harbor: Cold Spring Harbor Laboratory Press)
11. Small, J. A., Scangos, G. A., Cork, L., Jay, G. and Khoury, G. (1986). The early region of human papovirus JC induces dysmyelination in transgenic mice. *Cell*, **46**, 13-18
12. Gordon, J. W. and Ruddle, F. H. (1983). Gene transfer into mouse embryos: production of transgenic mice by pronuclear injection. *Methods Enzymol.*, **101**, 411-33
13. Small, J. A., Khoury, G., Jay, G., Hawley, P. M. and Scangos, G. A. (1986). Early regions of JC virus and BK virus induce distinct and tissue-specific tumors in transgenic mice. *Proc. Natl. Acad. Sci. USA*, **83**, 8288-92
14. Trapp, B. D., Itoyama, Y., MacIntosh, T. D. and Quarles, R. H. (1983). P_2 protein in oligodendrocytes and myelin of the rabbit central nervous system. *J. Neurochem*, **40**, 47-54
15. Trapp, B. D., Moench, T., Pulley, M., Barbosa, E., Tennekoon, G. and Griffin, J. (1987). Spatial segregation of mRNA encoding myelin-specific proteins. *Proc. Natl. Acad. Sci. USA*, **84**, 7773-7
16. Trapp, B. D., Small, J. A., Pulley, M., Khoury, G. and Scangos, G. A. (1988). Dysmyelination in transgenic mice containing JC virus early region. *Ann. Neurol.*, **23**, 38-48
17. Kenney, S., Natarajan, V., Strike, D., Khoury, G. and Salzman, N. P. (1984). JC virus enhancer-promoter active in human brain cells. *Science*, **226**, 1337-9
18. Lees, M. B. and Brostaff, S. W. (1984). Proteins of myelin. In Morell, P. (ed.) *Myelin*. 2nd edn., pp. 197-224. (New York: Plenum Press)
19. Quarles, R. H. (1983/1984). Myelin-associated glycoprotein in development and disease. *Dev. Neurosci.*, **6**, 285-303

20. Eng, L. F. (1980). The glial fibrillary acidic (GFA) protein. In Bradshaw, R. A. and Schneider, D. (eds.) *Proteins of the Nervous System*. pp. 85–117. (New York: Raven Press)
21. Hogan, E. L. and Greenfield, S. (1982). Animal models of genetic disorders of myelin. In Morell, P. (ed.) *Myelin*. 2nd edn., pp. 489–534. (New York: Plenum Press)
22. DePamphilis, M. L. and Wassarman, P. M. (1982). Organization and replication of papovavirus DNA. In Kaplan, A. S. (ed.) *Organization and Replication of Viral DNA*. pp. 37–114. (Boca Raton: CRS Press Inc.)
23. Rigby, P. W. J. and Land, P. D. (1983). Structure and function of Simian virus-40 large T-antigen. *Adv. Virol. Oncol.*, **3**, 31–47
24. Huber, B., Vakalopaulou, E., Burger, C. and Fanning, E. (1985). Identification and biochemical analysis of DNA replication-defective large T-antigens from SV40-transformed cells. *Virology*, **146**, 188–202
25. Cole, C. N., Tornow, J., Clark, R. and Tjian, R. (1986). Properties of the simian virus 40 (SV40) large T-antigens encoded by SV40 mutants with deletions in gene A. *J. Virol.*, **57**, 539–46
26. Stahl, H., Droge, P. and Knippers, R. (1986). DNA helicase activity of SV40 large tumor antigen. *EMBO J.*, **5**, 1939–44

4
Bovine Leukaemia: Facts and Hypotheses Derived from the Study of an Infectious Cancer

A. BURNY, Y. CLEUTER, R. KETTMANN, M. MAMMERICKX, G. MARBAIX, D. PORTETELLE, A. VAN DEN BROEKE, L. WILLEMS AND R. THOMAS

INTRODUCTION

Bovine leukaemia virus (BLV) is the aetiological agent of a chronic lymphatic leukaemia/lymphoma in cows, sheep and goats. Infection without neoplastic transformation was also obtained in pigs, rhesus monkeys, chimpanzees, rabbits and observed in capybaras and water-buffaloes. Structurally and functionally, BLV is a relative of human T-lymphotropic viruses 1 and 2 (HTLV-I and HTLV-II). HTLV-I induces in humans a T-cell leukaemia and its type 2 counterpart has been found in dermatopathic lymphadenopathy, hairy T-cell leukaemia and prolymphocytic leukaemia cases. At variance with HTLV-I, BLV has not been associated with neurological diseases of the degenerative type.

BLV, HTLV-I and HTLV-II show clearcut sequence homologies. The pathology of the BLV-induced disease, most notably the absence of chronic viraemia, a long latency period and lack of preferred proviral integration sites in tumours, is similar to that of adult T-cell leukaemia/lymphoma induced by HTLV-I. The most striking feature of the three naturally transmitted leukaemia viruses is the X region located between the *env* gene and the long terminal repeat (LTR) sequence. The X region contains several overlapping long open reading frames. One of them, designated XBL-I, encodes a *trans*-activator function capable of increasing the level of gene expression directed by BLV-LTR and most probably involved in 'genetic instability' of BLV-infected cells of the B-cell lineage. The genetic instability puts the cell into a context of fragility, ready to move through a number of stages towards full malignancy. Little is known about these events and their causes; we present some theoretical possibilities.

BLV infection has a worldwide distribution. In temperate climates the virus spreads mostly via iatrogenic transfer of infected lymphocytes. In warm climates and in areas heavily populated by haematophageous insects, there are indications of insect-borne propagation of the virus.

BLV GENOME AND GENE PRODUCTS

Bovine leukaemia (lymphoma, lymphosarcoma) is a contagious disease induced by bovine leukaemia virus (BLV), a retrovirus exogenous to the bovine species. It is a chronic disease, evolving over extended periods (1–8 years), with tumours developing in only a small number of infected animals. The same virus infects sheep where it induces tumours with very high frequency[1].

The BLV proviral genomic structure (Figure 4.1) has been established in detail by several authors[2-6]. Its salient features are:

(1)   the gag polyprotein contains virus structural proteins p15, p24 and p12;

(2)   a protease, p14, is coded by an open reading frame overlapping the *gag* gene on the left and the *pol* gene to the right; *gag*, protease (*prot*) and *pol* genes are in three different reading frames;

(3)   the *env* gene codes for a 72 000 env precursor (Pr 72^{env}) that is cleaved into two glycosylated envelope proteins gp51 and gp30;

(4)   two overlapping open reading frames, located between *env* and the 3' LTR code for a 34 kDa and a 16 kDa protein, respectively.

Two gag polyprotein precursors, 66 kDa (70 kDa) and 44 kDa are synthesized in BLV-infected cells and in reticulocyte cell-free lysates, programmed with BLV 38 S RNA and in frog oocytes micro-injected with the same RNA message[7-11]. The 66 kDa is the precursor to:

(1)   p15, myristilated fragment derived from the amino terminus;

(2)   p24, the major core protein;

(3)   p12, the RNA interacting fragment;

(4)   p14, the protease, probably synthesized via a frameshift suppression by a lysine-specific tRNA of the gag terminal codon[11].

The 44 kDa precursor lacks the p14 protease. P10 is an additional cleavage product found in purified BLV preparations; it is an amino terminal

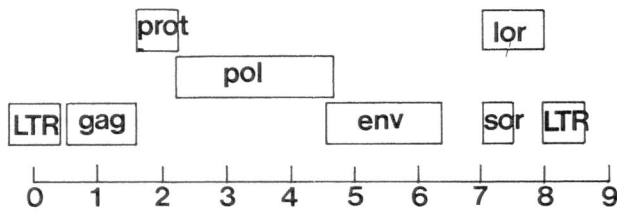

Figure 4.1   Genomic structure of BLV provirus

polypeptide of gag polyprotein. It is myristilated as well as p15, the larger and more common antigen derived from that region of the precursor.

The reverse transcriptase-endonuclease protein is made as a 145 kDa gag–pol precursor. Pr 145 indeed contains all the tryptic peptides of Pr 70^{gag}. It is present[8] in BLV-infected cells and in heterologous protein synthesizing systems programmed with BLV 38 S RNA as an elongation product of Pr 70^{gag}. No information is available about the mode of synthesis and cleavage of the gag–pol precursor which in BLV, might require continuous transcription from genetic information located in three different reading frames.

Pr 72^{env} is the precursor to gp51, the external envelope glycoprotein, and gp30, the transmembrane component. Pr 72^{env} overlaps the right hand side of pol by 17 amino acids and is cleaved into its two components at the Arg-Val-Arg-Arg sequence at the carboxyl end of gp51. The unglycosylated homologue of Pr 72^{env} is a 47 kDa polypeptide, which is not further processed. Mouse monoclonal antibodies to gp51 allowed the distinguishing of eight different epitopes, three of which (F, G and H) are associated with the biological activities of the virus (infectivity and syncytia induction). Processing of Pr 72^{env} and biological activity are conformation-dependent glycosylation-dependent phenomena. The neutralizing antibody-inducing sites of gp51 are subject to antigenic variations among BLV isolates of the same or different geographical origins[10,12,13]. One, at least, of the three presently identified epitopes (F, G and H) was always present on the BLV isolate. Moreover, no variation was found in the number and position of glycosylation sites and cysteine residues, stressing the uniformity of selection pressures exercised on BLV in all possible situations. Limited proteolytic hydrolysis of gp51 indicated that F, G and H epitopes are located in the amino-moiety of the protein (D. Portetelle, personal communication).

The transmembrane glycoprotein gp30 is a highly glycosylated 214 amino acid polypeptide. Of its six cysteine residues, four are conserved in all type C viruses, suggesting an invariant and crucial pattern of disulphide bonding. Gp30 anchors the envelope proteins in the membrane of the infected cell and the virus particle. It is not known whether S–S bridges between gp51 and gp30 serve as linkages between the two proteins after proteolytic cleavage of the precursor or whether there are artefacts. The easy loss of gp51 from virions during purification procedures seems to favour the latter part of the hypothesis[2,3,4,6,14]. Nearest neighbour relationships between components of BLV virions were investigated with chemical cross-linking and allowed the design of a structural model of the virus particle[15].

BLV genome contains two overlapping open reading frames 3′ to the envelope gene. The long open reading frame (lor or XBL-I) encodes a 34 kDa protein acting as transactivator of transcription of the provirus[16–20] and presumably of some cellular genes, which is believed to be the key process to initiation of cell transformation. The short open reading frame (sor or XBL-II) encodes a 16 kDa protein of presently undefined biological activity and most probably identical to the p16–p18 detected by Ghysdael et al.[7,9] by in vitro translation of BLV genomic RNA or poly A⁺ subgenomic fragments of it. P34 and p16 non-structural BLV proteins are supposedly translated from the same double spliced 2,1 kb mRNA[6,21,22]. Both proteins are recognized by sera of cattle in the tumour phase[23,24].

The genome structure, the nucleotide sequence of the provirus and the size and amino acid sequence of structural and non-structural viral proteins make BLV an obvious relative of human T-lymphotropic virus I and II and of simian T-lymphotropic virus III[2–6,16–20,24–30]. The four viruses replicate and act via transactivation of the enhancer–promoter regions of their LTRs and of cellular genes, and induce diseases with similar pathologies, most notably the absence of chronic viraemia, a long latency period and a lack of preferred proviral sites of integration in tumours.

It remains to be seen whether the full coding capacity of BLV DNA has been apprehended. It may be of interest to note that careful scrutiny of BLV cDNA libraries derived from virus-infected cells contain cDNA inserts that either correspond to partially spliced mRNAs or to messengers with presently uncovered physiological significance.

BLV INFECTION

Transmission

Transmission of BLV infection has been the subject of many field observations and experimental trials. Cases of natural infection are documented in cattle, sheep, capybara and water-buffalo (for review, see references 31 and 32). The infection can be experimentally transmitted to goats[33,34], pigs[35], rabbits (reference 36 and C. Altaner, personal communication), rhesus monkeys[37], chimpanzees[38] and buffaloes[39]. It has been amply documented that horizontal transmission is the rule including the transplacental route which amounts to 15% of infections in the offspring of BLV positive dams[40–43] and that infected cells are the best potential vehicles of infectious BLV particles. Consequently, the concentration of BLV-infected cells in the transmitted fluid (blood in most cases) is expected to play a major role in the success or failure of BLV transmission.

The data presented in Table 4.1 illustrate the outcome of an experiment carried out to determine the infectious dose of blood from a donor with persistent lymphocytosis (PL)[44]. Recipients were individual sheep labelled M125 to M135. Dilutions of whole blood from donor cow A (18 537 lymphocytes per µl of blood) were injected into each recipient and seroconversion was followed with time as a marker of infection. It appears that there is a threshold amount of blood to be inoculated in order to achieve successful infection. The last infectious dose corresponded to 926 lymphocytes. Of interest too was the observation that seroconversion was delayed from day 7 to day 18 post infection when the infectious load varied from $18\,537 \times 10^3$ to 926 lymphocytes. At 49 days post infection, antibody levels were all equal, irrespective of the number of lymphocytes administered. When the same cow was used as donor and goats as recipients, 1036 lymphocytes was the lowest dose leading to infection, an amount very similar to that required in sheep. Only twice that amount was required to achieve infection in adult cows, animals ten times heavier than sheep or goats.

40

Table 4.1 ELISA (gp BLV) titres in sera of 10 sheep inoculated with lymphocytes from a donor cow with persistent lymphocytosis

Animal	Number of lymphocytes inoculated	Days after inoculation											
		0	7	11	14	18	21	25	35	49	140	311	492
125	18 537 000	0	20	60	540	1 620	540	1 620	43 740	43 740	43 740	43 740	43 740
126	1 853 700	0	0	20	180	4 860	14 580	14 580	14 580	43 740	43 740	131 220	131 220
127	185 376	0	0	20	60	540	1 620	4 860	14 580	43 740	43 740	43 740	131 220
128	18 537	0	0	20	20	60	60	540	43 740	43 740	43 740	53 740	131 220
129	9 268	0	0	0	60	180	540	4 860	14 580	43 740	43 740	43 740	43 740
131	1 853	0	0	0	20	0	20	60	14 580	43 740	14 580	14 580	14 580
132	926	0	0	0	0	60	180	4 860	14 580	43 740	43 740	43 740	131 220
133	185	0	0	0	0	0	0	0	0	0	0	0	0
134	18	0	0	0	0	0	0	0	0	0	0	0	0
135	1	0	0	0	0	0	0	0	0	0	0	0	0

Taken from Mammerickx *et al.*, 1986; reference 44

In contrast, when the donor cow did not show persistent lymphocytosis, only the highest amount of blood (1 ml) established infection in recipient calves.

These data shed light on an important aspect of BLV transmission. Only animals in PL can be considered as highly infectious centres since minute amounts of their blood (50 nl in the documented example) transfer infection to naïve sheep or goats.

Repeated observations have indicated that, in temperate climates, the rate of BLV transmission parallels the intensity of veterinary care that is dispensed to a given herd. Considering the minute amount of blood that achieved infection from PL blood (reference 45 and see above), it is expected that the required dose can be transmitted via needles[46,47], dehorning gouges[43], etc. Successful infection can also occur via PL blood-contaminated gloves used in gynaecological examinations of cattle via the colorectal route (J. Everman, personal communication) emphasizing the high permeability of the colorectal mucosa to BLV-infected lymphocytes as has already been inferred for HIV transmission in human beings.

Identification of BLV-infected wild animals in tropical regions drew attention to the possible involvement of biting insects in the transmission of BLV. Observations currently made in the southwestern part of France (Les Landes) show a positive correlation between increase of BLV-infection rate in cattle and the number of tabanids captured (A. Parodi, personal communication). Experimental transmission of BLV from a PL cow to lambs via tabanid flies has been described[48]. Similar results were obtained by American investigators in Louisiana (L. D. Foil and C. J. Issel, personal communication). The latter authors could demonstrate transmission from a PL donor cow to sheep and goats with groups of 100 and 50 but not by groups of 25 and 10 horse flies (*Tabanus fuscicostatus*). In contrast, transmission failed when the donor was a non-PL animal.

Almost ironically, the above-mentioned observations give a sound epidemiological background to the old practice of eliminating animals in PL in attempts to combat bovine leukaemia at the herd level. The same observations may also be relevant to the possible transmission of HTLV-I via insect vectors, at least in some conditions prevalent in tropical areas[49,50]. Miller *et al.*[50] point out that 'Perhaps HTLV-I is only effectively transmitted by blood-sucking insects that have had an interrupted meal on a seropositive individual with some degree of lymphocytosis. This might then explain the apparent need for close familial contact over a prolonged period'. The observations made in the BLV system give direct support to this hypothesis.

Of utmost practical importance is the repeated observation that zona pellucida-intact bovine embryos can be transferred from bovine leukaemia virus-infected donors, including those bred by BLV-infected bulls[51], without risk of transmitting BLV, provided that they are properly washed prior to transfer[51-55]. As a result, elimination of BLV infection is feasible even in herds with very high genetic value where culling of infected recipients would be impractical due to the high breeding potential of the animal and thus the financial loss.

Protection against infection

As discussed above, it is assumed (and demonstrated in a number of cases) that BLV transmission results from the transfer of infected cells rather than free virus. The question thus arises whether it is feasible to achieve protection via vaccination procedures against a cell-linked pathogenic agent. The status of the question has been discussed by Miller[56]. From her survey of the known vaccination trials, it is obvious that very encouraging data have been obtained together with desperately poor ones. We remain, however, very optimistic about the design of an efficient BLV vaccine when considering that sheep passively immunized with anti-BLV antibody can successfully resist a challenge provided they have high enough anti-gp51 antibody titres[35,57]. This probably means that animals able to react with efficacy against the crucial determinants of BLV gp51 do not succumb to the challenge. We have shown by inhibition of syncytia induction and neutralization of pseudotypes that the three epitopes F, G and H, identified on gp51 by monoclonal antibodies, represent major determinants involved in the biological properties of the virus[10,12,13]. Furthermore, work performed in our group has amply demonstrated that F, G and H are conformational epitopes, glycosylation-dependent, and easily denatured when handling of gp51 is carried out without the utmost care. It follows that vaccination protocols that make use of native, biologically important, determinants of gp51 constitute a pre-requisite for a BLV vaccine. Experiments along these lines are under way.

HOST–VIRUS INTERPLAY: THE BLV TARGET CELL

Ab(+) animals with no lymphocytosis

In this category of Ab(+) animals the number of peripheral B lymphocytes increases while the total number of circulating lymphocytes is within the normal range.

Bovine leukaemia virus particles were first observed by Miller et al.[58] in short-term cultures of peripheral blood lymphocytes of BLV(+) animals in persistent lymphocytosis (see below). Numerous attempts to observe the virus in body fluids of PL animals or in tumours supposed to be of the enzootic bovine leukosis type had failed before and contributed not a little to making bovine leukaemia a puzzling disease with obvious infectious behaviour but without any observable agent. We now know that a viraemia can only be identified in the first 10 to 12 days post-infection preceding the appearance and the permanency from then on of anti-virus neutralizing antibodies[59].

As the mode of action of BLV now unfolds, it becomes apparent that mass production of virus particles in the host is not required. Very many sites of integration of the provirus in the host DNA allow expression of p34, the *trans*-activating protein. In some rare cases, the latter will make the target cell conducive to full malignant transformation. If massive production of BLV is not a prerequisite, then the more permissive the B-lymphocyte is to BLV, the better chance there is of the animal developing a tumour.

Permissivity of sheep cells to BLV and sensitivity of the species to BLV-induced tumours make indeed a striking parallel.

The permanency of anti-BLV antibody proves the existence of a permanent antigenic stimulation, mediated via viral proteins and particles produced by lymphocytes of the B-lineage and perhaps of other cell types. In fact, little is known about the identity of cells involved in BLV replication *in vivo*. It is established that BLV persists in peripheral B-lymphocytes[60] and that the proportion of B-lymphocytes in the peripheral blood of BLV(+) animals increases very significantly before any detectable increase in the number of circulating lymphocytes (C. Fossum, personal communication; H. A. Lewin, personal communication). The PL stage is thus preceded by the polyclonal stimulation of B cells and the subsequent increase in the proportion of B cells in the blood compartment of an apparently haematologically normal animal. Moreover, refined studies with monoclonal antibodies and fluorescence-activated cell sorting have shown that:

(1) the surface of some of the B cells in BLV-infected animals have altered sugar composition and seem to be arrested at an early stage of maturation;

(2) a small percentage of the Ig-bearing cells of BLV(+) animals are enlarged;

(3) an increased intensity of the fluorescence was observed both with the fluorescence microscope and by flow cytometry on a substantial proportion of the B-lymphocytes from BLV(+) cattle compared to B-lymphocytes from healthy cattle.

Such observations suggest that BLV might influence the regulation of Ig expression (C. Fossum, personal communication). Analogous data have been obtained by Lewin and co-workers (H. A. Lewin, personal communication).

The BLV genome is transcriptionally repressed in both transformed and non-transformed lymphocytes *in vivo*[61-63]. When isolated from their host and maintained in short-term cultures, PL lymphocytes and, in some cases, tumour cells (see below) express viral RNA and proteins. The mechanism by which BLV proviral transcription is repressed *in vivo* in B-lymphocytes is not understood. It apparently involves a plasma blocking factor (PBF)[64,65], a non-immunoglobulin plasma protein, found primarily but not solely in BLV-infected cattle. A platelet-derived factor (PDF) has an inhibitory effect on PBF[66] and exhibits high blastogenic activity on bovine lymphocytes. PDF was found in four fractions (fractions 1 to 4) corresponding to various molecular sizes after chromatography on Biogel 150. The most active fraction (fraction 4; 50 000 kDa) had physical and chemical properties similar to those of platelet-derived growth factor (PDGF). PDF and PBF acting antagonistically might be the main regulators of BLV expression in peripheral B-lymphocytes, a situation at variance with that prevailing in the HTLV-I system where expression of the integrated provirus is controlled by antibodies to envelope glycoprotein gp51[67].

44

Ab(+) animals with persistent lymphocytosis

Persistent lymphocytosis (PL) is a polyclonal proliferation of B cells occurring in some animals in response to constant stimulation by BLV antigens. As a consequence, the B/T cell ratio is modified concurrently with an increase (or not) of the total number of circulating lymphocytes. An animal is considered to be in PL when its total lymphocyte count significantly exceeds the value recognized as normal for animals of its class of age in several successive blood tests[68,69]. As discussed here above, recent data (C. Fossum, personal communication; H. A. Lewin, personal communication) have demonstrated that, in fact, PL prolongs and expands the phenomenon of B cell proliferation already evident in some animals in the preceding stage of BLV infection. As expected, unusual traits of B-lymphocytes already seen in some BLV-infected animals, such as modification of the glycosylation pattern of membrane components, increase in size, increased number of surface IgM molecules, etc., are evident when PL lymphocytes are examined by flow cytometry.

Earlier studies on the heritability of susceptibility to PL led to the conclusion that PL aggregates in some families[70]. Recent observations made in a herd of Shorthorn cattle ($n = 117$) suggested that the bovine lymphocyte antigen (BoLA) system plays a role in determining susceptibility or resistance to B cell proliferation and lymphocytosis. Relative resistance was associated with BoLA W7 whereas susceptibility was associated with BoLA W12.3 and BoLA W8[71]. The authors concluded that subclinical progression of BLV infection is under the control of the BoLA complex and suggested that the BoLA system can be used to select for resistance to B-cell proliferation and development of lymphocytosis in BLV-infected herds. They also noted that the same BoLA alleles are involved with BLV-related traits in different herds within a breed but may differ between breeds. Considering that PL animals are the major source of infection within a herd, it is highly advisable to eliminate the most sensitive haplotypes as a first step to reducing the overall rate of infection in heavily contaminated cattle populations.

Biochemical studies of numerous PL cases have shown that BLV provirus integrates at many possible sites in the cell genome. In PL, many different clones of BLV-infected cells proliferate; their propagation seems to be controlled at some times and seems to escape control at other times. PL can indeed remain stable for considerable periods of time but it can also become progressively more severe or suddenly disappear with restoration of a normal white blood cell count. No tumour cell clone seems to be present among PL lymphocytes; cell culture techniques that are successful in getting tumour cells to grow *in vitro* have never led to establishment of a cell line when used to PL lymphocytes[72] (also Y. Cleuter, personal communication).

Identification of PL lymphocytes has been the object of a few studies. PL cells contain both BLV-infected cells, at a resting stage, incorporating very little radioactive thymidine, and BLV-free, reactive cells, avidly incorporating the DNA precursor when stimulated *in vitro*[73,74]. As mentioned above, the vast majority of PL lymphocytes are B cells, 3–10% of which are large blastoid cells which represents a 30-fold increase compared to the proportion found in control animals (C. Fossum, personal communication). Flow cytometry analysis of PL lymphocytes labelled with fluoresceinated antibodies to bovine IgM showed a strong elevation of fluorescence intensity relative to that encountered with lymphocytes of normal cattle, indicating an increased number of surface immunoglobulin molecules (C. Fossum, personal communication; H. A. Lewin, personal communication). It was also observed that PL lymphocytes express an altered sugar composition on their surfaces compared to non-infected animals (C. Fossum, personal communication) and that induction of BLV, by short-term culture, may influence the regulation of Ig and BoLA class II antigen expression (H. A. Lewin, personal communication). Molecular hybridization data[75] have shown that the ratio of BLV(+) cells to BLV(-) cells (reactive cells) is roughly 1/3 to 1/4, a value in good agreement with the results of Kenyon and Piper[73,74].

In conclusion, PL or, more accurately, B cell proliferation with or without persistent lymphocytosis is a genetically controlled manifestation of BLV infection. The expanding B cell pool is heterogeneous with regard to the presence of BLV provirus and the stage of differentiation of the cells. Disregulation in the expression of sIg and sometimes BoLA class II antigen is obvious.

Ab(+) animals which develop tumours

Susceptibility to tumour development varies among the three ruminant species, cattle, sheep and goat. Sheep are the most sensitive. In one of our experimental trials, all infected animals succumbed from 6 months to 6 years post infection. Goats are the least sensitive, only two animals out of 20 developed a tumour within 10 years of BLV infection. Susceptibility to tumour development in BLV-infected cattle can be described as intermediate. Even when kept to old age, only a fraction, probably less than 10%, of BLV carriers will come down with a tumour. Irrespective of the animal species, tumours are lymphoid and involve the lymphatic system (tumorous lymph nodes) or are made of lymphoid masses of cells invading other tissues.

Liquid hybridization and Southern blot analysis using BLV-genome-specific probes showed that all BLV-induced tumours and most of the cells in individual tumours contain proviral genetic information, thus demonstrating the indispensible presence of the virus for tumour development[34,62,75]. Up to four copies of proviral DNA can be found per tumour cell, amongst these about 1/4 of the copies harbour deletions. We even observed tumours with a single deleted copy of the provirus per cell. In this case, however, the deletion does not cover the 5'LTR nor the 3' half of the provirus. We recently cloned and sequenced such a provirus. The deletion expanded over more than 4 kb from nucleotide 1030 in the p24 gene up to nucleotide 5332 in the middle of the gp51 gene, thus eliminating the spliced-in fragment at the end of the *pol* gene conserved in the *trans*-activation of transcription (tat) mRNA of *trans*-activating viruses (BLV, HTLV-I and II). Is the deleted provirus able to code for a functional tat-BLV protein? We tend to believe it can but have no proof so far. Our interpretation is that deletions occur at the provirus integration and can only be selected for by proliferation of the tumour clone if the deleted provirus expresses a functional tat-BLV protein. Expression of a functional tat protein is the mandatory prerequisite for BLV provirus to act as a tumour inducer and to replicate.

Early experiments[75] showed a striking difference between PL cells and tumours. In persistent lymphocytosis, proviral DNA was found to be integrated at a large number of genomic sites in one fourth to one third of circulating leukocytes. In the tumour form, in contrast, proviral DNA was found to be integrated at one or very few sites in the genome of the target cell. Tumours are the results of a mono- or oligoclonal proliferation of cells, these terms referring to the site(s) of BLV integration. Integration sites, however, are not conserved from one animal to the other. DNAs from 25 independent hamster × bovine somatic cell hybrids were submitted to the Southern blot analysis with probes made of unique cell DNA fragments adjacent to single-copy proviruses from three different bovine tumours. It appeared that these cellular sequences, and thus the respective proviruses, belonged to three different chromosomes in the three different tumours examined[76]. Previous experiments had, indeed, been designed to define large restriction fragments in normal cell DNA (15 and 17 kb, respectively) corresponding to the cellular domains surrounding two proviruses in the original tumours. No rearrangement of these two cellular loci, due to the insertion of a BLV provirus, was found in 28 other BLV-induced tumours[77]. We can thus safely conclude that the tumour cell can accommodate the proviral DNA sequences at many sites in the genome. This conclusion is well in line with the idea that transformation via BLV requires expression at least in the inducing phase of BLV tat protein. This requirement will be satisfied provided that BLV provirus resides in a region of open chromatin and presents the right signals and conformation for active transcription.

Studies of BLV expression in virus-induced tumours has led systematically to puzzling negative results. Tumour tissue taken *in vivo* does not express viral RNA nor viral proteins. It can be argued, however, that BLV-induced tumours being made essentially of resting cells, it was not surprising to find that the viral genes were silent. To circumvent the argument, we tried and succeeded to grow *in vitro* the tumour cells from four bovine and two ovine tumour cases. We could easily show[72] (also Y. Cleuter, personal communication) via the restriction profile of the proviral sequences, that the cells growing *in vitro*, were indeed the tumour cells. The restriction profiles were identical in the cultured cells and in the tumours they derived from. Among the six tumour cell lines thus developed, two express very little viral antigen (their viral RNA content amounts to about 0.15 molecule per cell) and four have no detectable viral RNA nor viral proteins (as assessed by Western blots with high affinity polyclonal antibody to BLV structural proteins). We are thus led to the apparent paradox that BLV is strictly mandatory as an inducer of the neoplastic process but is dispensable once the process has been switched on.

Little is known so far about the exact identity of the BLV-induced tumour cell. Flow cytometry analyses of our six tumour cell lines showed that they share the same phenotypic traits irrespective of their bovine or ovine origin. In short, they reacted with monoclonal antibodies to BoLA class I and II determinants and to heavy chains of IgM. No reactivity was encountered with

47

monoclonals specific for T or B cells or monocytes. Fluorescein-labelled antibodies to Ig light chains did not produce any signal above background (Y. Cleuter, personal communication). We tentatively concluded from these preliminary experiments that BLV-induced tumour cells, whether of bovine or ovine origin and propagated *in vitro*, exhibit characteristics of pre-B cells. More tumour cell-lines and more antibody reagents are being developed and, together with DNA probes to Ig genes, will allow more precise identification of the differentiation stage at which BLV-induced tumour cells are blocked.

FACTS AND HYPOTHESES ABOUT BLV AND CELL TRANSFORMATION

The mode of cell transformation by BLV, HTLV-I and HTLV-II remains conjectural. Beyond any doubt, the viral *tat* gene plays a key role within a given cellular background. The virus–cell interplay leads the cell to a given stage where it can stay for ever, being blocked in its differentiation pathway. The virus is certainly necessary but by no means sufficient. Given sets of circumstances which are rarely encountered, or rare secondary events, must contribute to pushing the cell across barriers to a state beyond which progress towards full malignancy is irreversible. We know nothing about the nature and number of such circumstances and events as may be involved in leading a cell to transformation in this way.

Considering that the presence of the virus is mandatory in the initial steps of the process and apparently dispensable later on, we hypothesize that the transient expression of viral functions can lead to permanent expression of critical cellular genes. Two possibilities can be considered: either the transformation process entirely rests upon regulatory mechanisms without alterations of DNA (sometimes called epigenetic processes) or else alterations of the genetic material of the cell make that cell susceptible to progress towards the neoplastic state. (This may include mutations, deletions, amplifications, translocations, visible chromosomal abnormalities, etc.). Obviously, in the BLV and HTLV-I and II systems, regulatory modifications are the results of *tat* expression which is turn does or does not induce alterations of host cell DNA.

Induction of cell transformation via regulatory pathways

As far as one can tell, a viral function, presumably *tat*, switches on host functions responsible for the development of tumours, and these host functions, once initiated, remain on even though all viral functions have been switched off. Two aspects will be briefly discussed:

trans-activation of host functions; and

the possible mechanisms by which a gene may be switched on by a transient signal and remain on after the signal has disappeared.

Trans-*activation of host functions*

That a regulatory gene can act in *trans* by emitting a diffusible regulator is one of the very bases of biological regulation uncovered by Jacob and Monod[78].

That these signals can be positive as well as negative has been shown by Englesberg[79] and by Thomas[80,81]. More specifically, *trans*-activation has been discovered in bacterial lysogens: silent prophage genes can be *trans*-activated (without lifting immunology) by superinfecting the lysogen with a related heteroimmune phage[80-82].

How can a transient signal switch a gene on permanently?

A transient signal might of course modify gene expression permanently by exerting an appropriate transmissible modification of DNA structure (see below). However, the same result can also be reached without any change at the level of DNA structure, provided the gene is (or is controlled by) a regulatory gene exerting a positive control on its own expression – directly or indirectly, via other genes forming with it a positive feedback loop. Let us first consider such a positive loop in the absence of the above-mentioned signal. Either the regulatory product is below its threshold of efficiency ('absent'), and since it is necessary to its own synthesis it will not be further synthesized and will remain indefinitely absent (positive loop 'off'); or the regulatory product is already above its threshold of efficiency ('present') and it will be further synthesized and remain indefinitely present (positive loop 'on'). Now, if the loop was initially off but it is switched on by a signal, it will remain on permanently even though the signal itself is transient. A concrete example is that of gene cI of bacteriophage. This gene is switched on by the positive regulator cII (signal); since cI exerts a positive control on its own synthesis, once activated (under cII control) it remains so (under its own control) even though cII expression is now repressed as well as other viral genes (by the cI product!)[83].

Many features of our system would be consistent with the idea that the transient expression of *tat* permanently switches on a positive loop made of host genes, of which one or more switch off the viral genes (including *tat*) and one or more are involved in the development of the tumour.

Induction of cell transformation via alterations of DNA

Abnormalities in the structure of chromosomes (translocations) or in their number (aneuploidy, hyperdiploidy, trisomy, etc.) are very frequently observed in human leukaemias and lymphomas[84]. Aneuploidy and structural modifications have also been reported in BLV-induced bovine leukaemia (for a review, see reference 31) and are observed in cultured tumour cells from cow and sheep in the tumour phase (P. Popescu, personal communication). The major question that arises is whether chromosomal aberrations are a primary event in tumorigenesis, leading to activation of oncogenes, of growth factor or of growth factor receptor genes, of differentiation genes, of metal ion regulating genes, etc.[85], or rather are secondary to the transformation process and reflect an adaptation of the cancer cells to tissue culture conditions or to abnormalities of metabolic pathways. Solving this problem is by no means a trivial issue.

Observations made in a number of myeloid leukaemia cases in man indicate that these diseases have a multi-step pathogenesis with clonal proliferation of marrow stem cells *preceding* acquisition of distinctive chromosomal, morphological and clinical abnormalities. The first step, as hypothesized by Fialkow and Singer[86] is characterized by genetic instability and confers a proliferative advantage on a cell clone. Other steps induce chromosomal abnormalities in the offspring of this clone and adaptation to abnormal metabolic conditions.

Chromosomal imbalances are also frequently observed in solid tumours[87] such as endometrial adenocarcinomas in man. They show several recurrent trisomies or tetrasomies: 1 (long arm), 10, 2, 7, 12, 3 and possibly X (long arm). These chromosomes or chromosome segments carry the majority of the genes coding for enzymes of glucidic metabolism, citrate cycle and initial steps of nucleotide synthesis[88]. These anomalies are interpreted as reflecting a disturbance of metabolic pathways whose enzymes are coded by housekeeping genes and thus an adaptive modification rather than a causative event. In favour of this interpretation is the repeated observation that no chromosomal anomalies are detected in the less advanced stages of endometrial carcinomas[89].

Aneuploidy and pseudodiploidy are also currently observed in B-cell lymphomas of baboons, and are thought to be induced by the concomitant infection of the animal by an HTLV-I-related, T-cell tropic virus and a herpes-type agent, which is B-cell tropic. Karyotypic anomalies are frequently observed in fresh lymphoma cells for a given set of chromosomes, namely chromosomes 20, 1, 4, 10, X which, interestingly enough, are the evolutional homologues of human chromosomes 14, 1, 6, 8, X which are also the most frequently involved in numerical and structural aberrations in human malignant lymphomas[90].

It is tempting to consider that BLV-*tat* gene product (p34) is the key molecule that introduces genetic instability into the cell to be transformed. P34 probably up-regulates or down-regulates expression of a number of genes. (For example, the high level of expression of BoLA class II molecules in our BLV-induced cultured tumour cells is striking.) As explained above, expression of *tat* may induce continuous expression of positively autoregulated cellular genes. Or *tat* expression may lead directly or indirectly to deregulation of expression of some critical genes such as histone genes, a condition known to initiate in yeast an imbalance of histone class proteins which, in turn, induces chromosomal abnormalities. Such karyotypic rearrangements may well markedly affect expression of some oncogenes, differentiation genes, metal-ion-regulating genes, growth factor and growth factor receptor genes, etc. Known examples are the *myc* and *abl* oncogenes in Burkitt lymphoma and Philadelphia chromosome-positive chronic myelogenous leukaemia, respectively. Putatively new oncogenes *bcl-1* and *bcl-2* are located on chromosomes 11q13 and 18q21, respectively, and are thought to be activated in B-cell lymphomas (for review see references 84 and 91). The abnormal ploidies exemplified here may also be consequences of *tat* expression.

We are thus led to conclude that *tat* expression is a primary event whose neoplastic effect may be mediated via genetic instability. The malignant state, the final stage in the multi-step process for B-cell lymphomagenesis[92] is maintained, we have argued, either by positive regulatory loops without karyotypic alterations or via permanent modifications of the genome materialized by chromosomal abnormalities. If this is so, it is understandable that no expression of the BLV provirus is required to maintain the transformed state. The proviral information plays a major role which is mandatory but not sufficient. As soon, however, as a critical step in the tumorigenic process has been reached, the provirus is of no use, it is dispensible. We do not know the identity of this critical step, nor do we appreciate the range of possibilities included in this step. We have, indeed, several cell clones that do not grow well *in vitro*. It is possible that a number of cells escape the leukaemic block, differentiate and die while others remain neoplastic and divide. Should this hypothesis be verified experimentally, it would show that there are degrees in the intensity of transformation. What are these degrees? What are the molecules involved and the regulatory circuits in which they play a role?

Karyotypic analysis of our six BLV-induced tumour cell-lines shows that profound rearrangement of one X chromosome is a common event (P. Popescu, personal communication). Comparisons between fresh, stimulated tumour cells and their cultured counterparts will demonstrate whether the chromosomal abnormalities pre-exist in the original tissue or not. If the chromosomal aberrations are the result of *in vitro* culture, this might suggest that BLV-induced tumours are conditioned neoplasms[93-95] progressing to autonomous behaviour via chromosome rearrangements.

The fruitful work performed in the last few years on *trans*-activating retroviruses is shedding light on the initial steps of cell transformation. The BLV, HTLV-I and HTLV-II systems should provide basic clues about the chain of events that lead T- and B-committed lymphocytes from normal to neoplastic behaviour.

Acknowledgements

The work performed in the authors' laboratory was helped financially by the *Fonds Cancérologique de la Caisse Générale d'Epargne et de Retraite* and by the Ministry of Agriculture. R. Kettmann and G. Marbaix are *Maître de Recherche* and L. Willems is *Aspirant* of the *Fonds National de la Recherche Scientifique*, A. Van den Broeke is a Fellow of the Lady Tata Memorial Trust.

References

1. Mammerickx, M., Dekegel, D., Burny, A. and Portetelle, D. (1976). Study on the oral transmission of bovine leukosis to the sheep. *Vet. Microbiol.*, **1**, 347-50
2. Rice, N. R., Stephens, R. M., Couez, D., Deschamps, J., Kettmann, R., Burny, A. and Gilden, R. V. (1984). The nucleotide sequence of the *env* gene and post-*env* region of bovine leukemia virus. *Virology*, **138**, 82-93
3. Rice, N. R., Stephens, R. M., Burny, A. and Gilden, R. V. (1985). The *gag* and *pol* genes of bovine leukemia virus: nucleotide sequence and analysis. *Virology*, **142**, 357-77
4. Sagata, N., Yasunaga, T., Tsuzuku-Kawamura, J., Ohishi, K., Ogawa, Y. and Ikawa, Y. (1985). Complete nucleotide sequence of the genome of bovine leukemia virus: its evolutionary relationship to other retroviruses. *Proc. Natl. Acad. Sci. USA*, **82**, 677-81

5. Weiss, R., Varmus, H. and Coffin, J. (eds.) (1984). *RNA Tumor Viruses*. (Cold Spring Harbor, N.Y.: Cold Spring Harbor Laboratory)
6. Rice, N. R., Stephens, R. M. and Gilden, R. V. (1987). Sequence analysis of the bovine leukemia virus genome. In Burny, A. and Mammerickx, M. (eds.) *Enzootic Bovine Leukosis and Bovine Leukemia Virus*. pp. 115–44. (Boston: Martinus Nijhoff)
7. Ghysdael, J., Kettmann, R. and Burny, A. (1978). Translation of bovine leukaemia virus genome information in heterologous protein synthesizing systems programmed with virion RNA and in cell-lines persistently infected by BLV. *Ann. Rech. Vet.*, **9**, 627–34
8. Ghysdael, J., Kettmann, R. and Burny, A. (1979). Translation of BLV virion RNAs in heterologous protein-synthesizing systems. *J. Virol.*, **29**, 1087–98
9. Ghysdael, J., Bruck, C., Kettmann, R. and Burny, A. (1984). Bovine leukemia virus. In Vogt, P. K. and Koprowski, H. (eds.) *Current Topics in Microbiology and Immunology*. pp. 1–19. (Berlin: Springer Verlag)
10. Bruck, C., Rensonnet, N., Portetelle, D., Cleuter, Y., Mammerickx, M., Burny, A., Mamoun, R., Guillemain, B., Van der Maaten, M. and Ghysdael, J. (1984). Biologically active epitopes of bovine leukemia virus glycoprotein gp51: their dependence on protein glycosylation and genetic variability. *Virology*, **136**, 20–31
11. Yoshinaka, Y., Katoh, I., Copeland, T., Smythers, G. W. and Oroszlan, S. (1986). Bovine leukemia virus protease: purification, chemical analysis and *in vitro* processing of gag precursor polyproteins. *J. Virol.*, **57**, 826–32
12. Bruck, C., Mathot, S., Portetelle, D., Berte, C., Franssen, J. D., Herion, P. and Burny, A. (1982). Monoclonal antibodies define eight independent antigenic regions on the bovine leukemia virus (BLV) envelope glycoprotein gp51. *Virology*, **122**, 342–52
13. Bruck, C., Portetelle, D., Burny, A. and Zavada, J. (1982). Topographical analysis by monoclonal antibodies of BLV gp51 epitopes involved in viral functions. *Virology*, **122**, 353–62
14. Schultz, A. M., Copeland, T. D. and Oroszlan, S. (1984). The envelope proteins of bovine leukemia virus: purification and sequence analysis. *Virology*, **135**, 417–27
15. Uckert, W., Wunderlich, V., Ghysdael, J., Portetelle, D. and Burny, A. (1984). Bovine leukemia virus (BLV) – A structure model based on chemical cross-linking studies. *Virology*, **133**, 386–92
16. Derse, D., Caradonna, S. J. and Casey, J. W. (1985). Bovine leukemia virus long terminal repeat: A cell type-specific promoter. *Science*, **227**, 317–20
17. Derse, D. and Casey, J. W. (1986). Two elements in the bovine leukemia virus long terminal repeat that regulate gene expression. *Science*, **231**, 1437–40
18. Rosen, C. A., Sodroski, J. G., Kettmann, R., Burny, A. and Haseltine, W. A. (1985). *Trans*-activation of the bovine leukemia virus long terminal repeat in BLV-infected cells. *Science*, **227**, 321–3
19. Rosen, C. A., Sodroski, J. G., Kettmann, R. and Haseltine, W. A. (1986). Activation of enhancer sequences in type II human T-cell leukemia virus and bovine leukemia virus long terminal repeats by virus-associated transacting regulatory factors. *J. Virol.*, **57**, 738–44
20. Rosen, C. A., Sodroski, J. G., Willems, L., Kettmann, R., Campbell, K., Zaya, R., Burny, A. and Haseltine, W. A. (1986). The 3′ region of bovine leukemia virus genome encodes a *trans*-activator protein. *EMBO J.*, **5**, 2585–9
21. Mamoun, R., Astier-Gin, T., Kettmann, R., Deschamps, J., Rebeyrotte, N. and Guillemain, B. (1985). The px region of the bovine leukemia virus is transcribed as a 2,1 kilobase mRNA. *J. Virol.*, **54**, 625–9
22. Sagata, N., Tsuzuku-Kawamura, J., Nagayoshi-Aida, M., Shimizu, F., Imagawa, K. I. and Ikawa, Y. (1985). Identification and some biochemical properties of the major *X BL* gene product of bovine leukemia virus. *Proc. Natl. Acad. Sci. USA*, **82**, 7879–83
23. Yoshinaka, Y. and Oroszlan, S. (1985). Bovine leukemia virus post-envelope gene coded protein: evidence for expression in natural infection. *Biochem. Biophys. Res. Commun.*, **131**, 347–54
24. Willems, L., Bruck, C., Portetelle, D., Burny, A. and Kettman, R. (1986). Expression of a cDNA clone corresponding to the long open reading frame XBL-I of the bovine leukemia virus. In Neth, R., Greaves, M. F. and Gallo, R. C. (eds.) *Modern Trends in Human Leukemia VIII*. (Berlin: Springer Verlag)
25. Poiesz, B. J., Ruscetti, F. W., Gazdar, A. F., Bunn, P. A., Minna, J. D. and Gallo, R. C.

(1980). Isolation of type C retrovirus particles from cultured and fresh lymphocytes of a patient with cutaneous T-cell lymphoma. *Proc. Natl. Acad. Sci. USA*, **77**, 7415-9

26. Kalyanaraman, V. S., Sarngadharan, M. G., Robert-Guroff, M., Miyoshi, I., Blayney, D., Golde, D. and Gallo, R. C. (1982). A new subtype of human T-cell leukemia virus (HTLV-II) associated with a T-cell variant of hairy cell leukemia. *Science*, **218**, 571-3

27. Seiki, M., Hattori, S., Hirayama, Y. and Yoshida, M. (1983). Human adult T-cell leukemia virus: complete nucleotide sequence of the provirus genome integrated in leukemia cell DNA. *Proc. Natl. Acad. Sci. USA*, **80**, 3618-22

28. Seiki, M., Hikikoshi, A., Taniguchi, T. and Yoshida, M. (1985). Expression of the *pX* gene of HTLV-I. General splicing mechanism in the HTLV family. *Science*, **288**, 1532-4

29. Sodroski, J. G., Rosen, C. A. and Haseltine, W. A. (1984). *Trans*-acting transcriptional activation of the human T lymphotropic virus long terminal repeat in infected cells. *Science*, **225**, 381-4

30. Oroszlan, S., Copeland, T. D., Henderson, L. E., Stephenson, J. R. and Gilden, R. V. (1979). Amino terminal sequence of bovine leukemia virus major internal protein: homology with mammalian type C virus p30s. *Proc. Natl. Acad. Sci. USA*, **76**, 2996-3000

31. Burny, A., Bruck, C., Chantrenne, H., Cleuter, Y., Dekegel, D., Ghysdael, J., Kettmann, R., Leclercq, M., Leunen, J., Mammerickx, M. and Portelelle, D. (1980). Bovine leukemia virus: molecular biology and epidemiology. In Klein, G. (ed.) *Viral Oncology*. pp. 231-89. (New York: Raven Press)

32. Marin, C., de Lopez, N., de Alvarez, L., Castanos, H., Espana, W., Leon, A. and Bello, A. (1982). Humoral spontaneous response to bovine leukemia virus infection in zebu, sheep, buffalo and capybara. *Curr. Top. Vet. Med. Anim. Sci.*, **15**, 310-20

33. Olson, C., Kettmann, R., Burny, A. and Kaja, R. (1981). Goat lymphosarcoma from bovine leukemia virus. *J. Natl. Cancer Inst.*, **67**, 671-5

34. Kettmann, R., Mammerickx, M., Portelelle, D., Grégoire, D., Burny, A. (1984). Experimental infection of sheep and goat with bovine leukemia virus: localization of proviral information in the target cells. *Leukemia Res.*, **8**, 937-44

35. Mammerickx, M., Portelelle, D. and Burny, A. (1981). Experimental cross-transmission of bovine leukemia virus (BLV) between several animal species. *Zentralbl. Veterinaermed. B.*, **28**, 69-81

36. Burny, A., Cleuter, Y., Couez, D., Dandoy, C., Gras-Masse, H., Gregoire, D., Kettmann, R., Mammerickx, M., Marbaix, G., Portelelle, D., Tartar, A., Van den Broeke, A. and Willems, L. (1985). Bovine leukemia virus (BLV) as a model system for human lymphotropic virus (HTLV) and HTLV as a model for BLV. In Deinhardt, F. (ed.) *Proceedings of the XIIth Symposium for Comparative Research on Leukemia and Related Diseases.* Hamburg, 7-11 July, 1985. pp. 336-48

37. Schödel, F., Hahn, B., Hübner, R. and Hochstein-Mintzel, V. (1986). Transmission of bovine leukemia virus (BLV) to immunocompromised monkeys: evidence for persistent infection. *Microbiologica*, **9**, 163-72

38. Van der Maaten, M. J. and Miller, J. M. (1976). Serological evidence of transmission of bovine leukemia virus to chimpanzees. *Vet. Microbiol.*, **1**, 351-7

39. Persechino, A., Montemagno, F. and D'Amore, L. (1984). Sulla recettivia del buffalo al virus della leucosi bovina enzootica. II. Prova di infezione sperimentale. *Att. Soc. Ital. Buiatr.*, **15**, 497-8

40. Piper, C. E., Abt, D. A., Ferrer, J. F. and Marshak, R. R. (1975). Seroepidemiological evidence for horizontal transmission of bovine. C-type virus. *Cancer Res.*, **35**, 2714-6

41. Piper, C. E., Ferrer, J. F., Abt, D. A. and Marshak, R. R. (1979). Postnatal and prenatal transmission of the bovine leukemia virus under natural conditions. *J. Natl. Cancer Inst.*, **62**, 165-8

42. Van der Maaten, M. J., Miller, J. M. and Schmerr, M. J. (1981). *In utero* transmission of bovine leukemia virus. *Am. J. Vet. Res.*, **42**, 1052-4

43. Di Giacomo, R. F., Darlington, R. L. and Evermann, J. F. (1984). Natural transmission of bovine leukaemia virus in dairy calves by dehorning. *Can. J. Comp. Med.*, **49**, 340-2

44. Mammerickx, M., Portelelle, D., De Clercq, C. and Burny, A. (1987). Experimental transmission of bovine leukemia virus to cattle, sheep and goats: infectious doses of blood and incubation period of the infection. *Leukemia Res.*, **11**, 353-8

45. Van der Maaten, M. J. and Miller, J. M. (1977). Susceptibility of cattle to bovine leukemia virus infection by various routes of exposure. In Bentvelzen, P., Hilgers, J. and Yohn, D. S. (eds.) *Advances in Comparative Leukemia Research*. pp. 29–32. (New York: Elsevier/ North Holland Biomedical Press)
46. Wilesmith, J. W. (1979). Needle transmission of bovine leukosis virus. *Vet. Rec.*, **104**, 107
47. Evermann, J. F., Di Giacomo, R. F., Ferrer, J. F. and Parish, S. M. (1986). Transmission of bovine leukosis virus by blood inoculation. *Am. J. Vet. Res.*, **47**, 1885–7
48. Oshima, K., Okada, K., Numakunai, S., Yoneyama, Y., Sato, S. and Takahashi, K. (1981). Evidence on horizontal transmission of bovine leukemia virus due to blood-sucking tabanid flies. *Jpn. J. Vet. Sci.*, **43**, 79–81
49. Greaves, M. F. and Miller, G. J. (1986). Are haematophagous insects vectors for HTLV-I? In Neth, R., Greaves, M. F. and Gallo, R. C. (eds.) *Modern Trends in Human Leukaemia VII*. (Berlin: Springer Verlag) (In press)
50. Miller, G. J., Pegram, S. M., Kirkwood, B. R., Beckles, G. L. A., Byam, N. T. A., Clayden, S. A., Kinlen, L. J., Chan, L. C., Carson, D. C. and Greaves, M. F. (1986). Ethnic composition, age, sex and the location and standard of housing as determinants of HTLV-I infection in an urban Trinidadian community. *Int. J. Cancer*, **38**, 801–8
51. Monke, D. R. (1986). Noninfectivity of semen from bulls infected with bovine leukosis virus. *J. Am. Vet. Med. Assoc.*, **188**, 823–6
52. Parodi, A. L., Manet, G., Vuillaume, A., Crespeau, F., Toma, B. and Levy, D. (1983). Transplantation embryonnaire et transmission de l'agent de la leucose bovine enzootique. *Bull. Acad. Vet. Fr.*, **56**, 183–9
53. Kaja, R. W., Olson, C., Rowe, R. F., Stauffacher, R. H., Strozinski, L. L., Hardie, A. R. and Bause, I. (1984). Establishment of a bovine leukosis virus-free dairy herd. *J. Am. Vet. Med. Assoc.*, **184**, 184–5
54. Hare, W. C. D., Mitchell, D. Singh, E. L., Bouillant, A. M. P., Eaglesome, M. D., Ruckerbauer, G. M., Bielanski, A. and Randall, G. C. B. (1985). Embryo transfer in relation to bovine leukemia virus control and eradication. *Can. Vet. J.*, **26**, 231–4
55. Di Giacomo, R. F., Studer, E., Evermann, J. F. and Evered, J. (1986). Embryo transfer and transmission of bovine leukosis virus in a dairy herd. *J. Am. Vet. Med. Assoc.*, **188**, 827–8
56. Miller, J. (1986). Bovine leukemia virus vaccine. In Salzman, L. A. (ed.) *Animal Models of Retrovirus Infection and their Relationship to AIDS*. pp. 421–30. (New York: Academic Press)
57. Kono, Y., Arai, K., Sentsui, H., Matsukawa, S. and Itohara, S. (1986). Protection against bovine leukemia virus infection in sheep by active and passive immunization. *Jpn. J. Vet. Sci.*, **48**, 117–25
58. Miller, J. M., Miller, L. D., Olson, C. and Gillette, K. G. (1969). Virus-like particles in phytohemagglutinin-stimulated lymphocyte cultures with reference to bovine lymphosarcoma. *J. Natl. Cancer Inst.*, **43**, 1297–305
59. Portetelle, D., Bruck, C., Burny, A., Dekegel, D., Mammerickx, M. and Urbain, J. (1978). Detection of complement-dependent lytic antibodies in sera from bovine leukemia virus-infected animals. *Ann. Rech. Vet.*, **9**, 667–74
60. Paul, P. S., Pomeroy, K. A., Johnson, D. W., Muscoplat, C. C., Handwerger, B. S., Soper, F. F. and Sorensen, D. K. (1977). Evidence for the replication of bovine leukemia virus in the B lymphocytes. *Am. J. Vet. Res.*, **38**, 873–6
61. Kettmann, R., Marbaix, G., Cleuter, Y., Portetelle, D., Mammerickx, M. and Burny, A. (1980). Genomic integration of bovine leukemia provirus and lack of viral RNA expression in the target cells of cattle with different responses to BLV infection. *Leukemia Res.*, **4**, 509–19
62. Kettmann, R., Deschamps, J., Cleuter, Y., Couez, D., Burny, A. and Marbaix, G. (1982). Leukemogenesis by bovine leukemia virus: proviral DNA integration and lack of RNA expression of viral long terminal repeat and 3' proximate cellular sequences. *Proc. Natl. Acad. Sci. USA*, **79**, 2465–9
63. Marbaix, G., Kettmann, R., Cleuter, Y. and Burny, A. (1981). Viral RNA content of bovine leukemia virus-infected cells. *Mol. Biol. Rep.*, **7**, 135–8
64. Gupta, P. and Ferrer, J. F. (1982). Expression of bovine leukemia virus genome is blocked by a nonimmunoglobulin protein in plasma from infected cattle. *Science*, **215**, 405–7

65. Gupta, P., Kashmiri, S. V. S. and Ferrer, J. F. (1984). Transcriptional control of the bovine leukemia virus genome: role and characterization of a non-immunoglobulin plasma protein from bovine leukemia virus-infected cattle. *J. Virol.*, **50**, 267–70
66. Tsukiyama, K. (1985). Control of expression of bovine leukemia virus genome: effect of plasma blocking factor and platelet-derived growth factor. *Jpn. J. Vet. Sci.*, **33**, 101
67. Tochikura, T., Iwahashi, M., Matsumoto, T., Koyanagi, Y., Hinuma, Y. and Yamamoto, N. (1985). Effect of human serum anti-HTLV antibodies on viral antigen induction in *in vitro* cultured peripheral lymphocytes from adult T-cell leukemia patients and healthy virus carriers. *Int. J. Cancer*, **36**, 1–7
68. Bendixen, H. J. (1963). *Leukosis enzootica bovis. Diagnostik, Epidemiologi, Bekaempelse.* (Copenhagen: Carl F. Mortensen)
69. Burny, A. and Mammerickx, M. (eds.) (1987). *Enzootic Bovine Leukosis and Bovine Leukemia Virus.* (Boston: Martinus Nijhoff)
70. Abt, D. A., Marshak, R. R., Kulp, H. W. and Pollock, R. J. Jr. (1970). Studies on the relationship between lymphocytosis and bovine leukosis. *Bibl. Haematol.*, **36**, 527–36
71. Lewin, H. A. and Bernoco, D. (1986). Evidence for BoLA-linked resistance and susceptibility to subclinical progression of bovine leukaemia virus infection. *Anim. Genet.*, **17**, 197–207
72. Kettmann R., Cleuter, Y., Gregoire, D. and Burny, A. (1985). Role of the 3′ long open reading frame region of bovine leukemia virus in the maintenance of cell transformation. *J. Virol.*, **54**, 899–901
73. Kenyon, S. J. and Piper, C. E. (1977). Cellular basis of persistent lymphocytosis in cattle infected with bovine leukemia virus. *Infect. Immun.*, **16**, 891–7
74. Kenyon, S. J. and Piper, C. E. (1977). Properties of density gradient-fractionated peripheral blood leukocytes from cattle infected with bovine leukemia virus. *Infect. Immun.*, **16**, 898–903
75. Kettmann, R., Cleuter, Y., Mammerickx, M., Meunier-Rotival, M., Bernardi, G., Burny, A. and Chantrenne, H. (1980). Genomic integration of bovine leukemia provirus: Comparison of persistent lymphocytosis with lymph node tumor form of enzootic bovine leukosis. *Proc. Natl. Acad. Sci. USA*, **77**, 2577–81
76. Gregoire, D., Couez, D., Deschamps, J., Heuertz, S., Hors-Cayla, M. C., Szpirer, J., Szpirer, C., Burny, A., Huez, G. and Kettmann, R. (1984). Different bovine leukemia virus-induced tumors harbor the provirus in different chromosomes. *J. Virol.*, **50**, 275–9
77. Kettmann, R., Deschamps, J., Couez, D., Claustriauz, J. J., Palm, R. and Burny, A. (1983). Chromosome integration domain for bovine leukemia provirus in tumors. *J. Virol.*, **47**, 146–50
78. Jacob, F. and Monod, J. (1961). Genetic regulatory mechanisms in the synthesis of proteins. *J. Mol. Biol.*, **3**, 318–56
79. Englesberg, E., Irr, J., Power, J. and Lee, N. (1965). Positive control of enzyme synthesis by gene C in the l-arabinose system *J. Bacteriol.*, **90**, 946–57
80. Thomas, R. (1966). Control of development in temperate bacteriophages. I. Induction of prophage genes following heteroimmune super infection. *J. Mol. Biol.*, **22**, 79–95
81. Thomas, R. (1970). Control of development in temperate bacteriophages. III. Which prophage genes are and which are not *trans*-activable in the presence of immunity? *J. Mol. Biol.*, **49**, 393–404
82. Dambly, C., Couturier, M. and Thomas, R. (1968). Control of development in temperate bacteriophages. II. Control of lysozyme synthesis. *J. Mol. Biol.*, **32**, 67–81
83. Eisen, H., Brachet, P., Pereira da Silva, L. and Jacob, F. (1970). Regulation of repressor expression in λ. *Proc. Natl. Acad. Sci. USA*, **66**, 855–62
84. Goldman, J. M. and Harnden, D. G. (eds.) (1986). *Genetic Rearrangements in Leukaemia and Lymphoma.* (London: Churchill Livingstone)
85. Van den Berghe, H. and Mecucci, C. (1986). Some karyotypic aspects of human leukemia. In Neth, R., Greaves, M. F. and Gallo, R. C. (eds.) *Modern Trends in Human Leukemia VII.* (Berlin: Springer Verlag) (In press)
86. Fialkow, P. J. and Singer, J. W. (1985). Tracing development and cell lineages in human hemopoietic neoplasia. In Weissman, I. L. (ed.) *Leukemia.* Dahlem Konferenzen, 1985. pp. 203–22. (Berlin: Springer Verlag)

87. Dutrillaux, B., Muleris, M. and Gerbault-Seureau, M. (1986). Imbalance of sex chromosomes, with gain of early-replicating X, in human solid tumors. *Int. J. Cancer*, **37**, 475-9
88. Dutrillaux, B. and Couturier, J. (1986). Chromosome imbalances in endometrial adenocarcinomas: a possible adaptation to abnormal metabolic pathways. *Ann. Genet.*, **29**, 76-81
89. Fujita, H., Wake, N., Kutsuzawa, T., Ichinoe, K., Hreschyshyn, M. M. and Sandberg, A. A. (1985). Marker chromosomes of the long arm of chromosome 1 in endometrial carcinoma. *Cancer Genet. Cytogenet.*, **18**, 283-93
90. Markaryan, D. S. and Popova, E. A. (1986). Cytogenetics of monkey malignant lymphomas. A comparison with cytogenetics of human malignant lymphomas. In *International conference on primary-localized haematopoietic tissue tumors of the non-human primates and the ways of their generalization*. Sukhumi, USSR. Abstract, p. 9
91. Klein, G. and Klein, E. (1985). *myc/Ig* juxtaposition by chromosomal translocations: some new insights, puzzles and paradoxes. *Immunol. Today*, **6**, 208-15
92. Gordon, J., Aman, P., Rosen, A., Ernberg, I., Ehlin-Henricksson, B. and Klein, G. (1985). Capacity of B-lymphocytic lines of diverse tumor origin to produce and respond to B-cell growth factors: a progression model for B-cell lymphomagenesis. *Int. J. Cancer*, **35**, 251-6
93. Furth, J. (1953). Conditioned and autonomous neoplasms: a review. *Cancer Res.*, **13**, 477-92
94. Foulds, L. (1958). The natural history of cancer. *J. Chronic Dis.*, **8**, 2-37
95. Klein, G. and Klein, E. (1986). Conditioned tumorigenicity of activated oncogenes. *Cancer Res.*, **46**, 3211-24
96. Olson, C., Rowe, R. F. and Kaja, R. (1982). Embryo transplantation and bovine leukosis virus: preliminary report. In Straub, O. C. (ed.) *Fourth International Symposium on Bovine Leukosis*, Bologna, 1980. pp. 361-9. (The Hague: Martinus Nijhoff)

5
Human Immunodeficiency Virus (HIV) Brain Infection in Infants and Children

L. G. EPSTEIN

INTRODUCTION

The acquired immunodeficiency syndrome (AIDS) and AIDS-related complex (ARC) are caused by the retrovirus now designated the human immunodeficiency virus (HIV)[1-4]. The spectrum of HIV infection in adults and children includes neurological dysfunction in the majority of patients[5-9]. It has been hypothesized that the progressive encepalopathy seen in children with HIV infection is due to direct brain infection with this retrovirus[5, 10-12]. Evidence in support of this hypothesis includes the transmission of HIV infection to chimpanzees using brain tissue[13], the isolation of this virus from brain and cerebrospinal fluid (CSF) of patients with AIDS and ARC[14-16], and the identification of HIV genome in the brain of adults and children with AIDS encephalopathy[12]. Further indirect evidence comes from analogies with animal retroviruses of the lentivirus subfamily, such as visna[17] and the newly described STLV-III in macaques[18], which are known to be neurotropic.

This chapter will first review some of the pertinent clinical and neuropathological findings of HIV brain infection in children. This will be followed by a brief discussion of recent studies of HIV antibodies and antigen expression in the CSF of children with progressive encephalopathy, and, finally, possible pathogenic mechanisms which may explain the clinical, CSF and neuropathological findings observed in children with HIV brain infection will be discussed.

CLINICAL FEATURES OF HIV BRAIN INFECTION IN CHILDREN

Children most often acquire HIV infection from their mothers who are infected with this retrovirus[19]. These women are usually intravenous drug abusers or sexual partners of HIV seropositive men. Less commonly children are infected via transfusion with HIV-contaminated blood products[19]. It is thought that both *in utero* and perinatal transmission of HIV infection from mother to child occur.

The time of HIV infection of brain is not known. In one report HIV was isolated from fetal tissues including brain at 20 weeks of gestation[20]. Clinical signs of neurological deterioration have been noted as early as 2 months and as late as 5 years after birth in children who acquired HIV infection from their mothers[7]. Neuropathological changes have been observed as early as 4 months after birth, but tended to be more prominent in children who died at a later age[11]. HIV-infected infants are usually born normocephalic and later become microcephalic[7]. In spite of one preliminary report[21], neither malformations of the brain nor a distinct embryopathy due to HIV infection have been convincingly demonstrated to date. Initial attempts to identify HIV genome in fetal brain tissue in two cases have thus far yielded only negative results (L. Epstein, unpublished data).

Children with HIV infection often manifest a progressive encephalopathy with features listed including: loss of developmental milestones in infants or subcortical dementia in the older children; impaired brain growth; progressive motor dysfunction with paresis, pathological reflexes, abnormal tone, pseudobulbar palsy, or ataxia; and less commonly seizures, myoclonus or extrapyramidal rigidity. The neurological deterioration occurs in a stepwise manner with plateaux of relative stability lasting months alternating with periods of more rapid decline occurring over several weeks. The end stage is characterized by mutism with spasticity in flexion.

Computed tomographic (CT) scans show varying degrees of cerebral atrophy and may, in addition, show enhancement or calcifications of the basal ganglia or periventricular white matter[7, 22, 23].

For the purpose of statistical analysis we classify our cohort of children with HIV infection into three categories descriptive of their neurological status: progressive encephalopathy, static encephalopathy, or normal neurological examination. This is based on the working hypothesis that the progressive encephalopathy which occurs in these children is due to primary retroviral infection of the brain. The population of children in which HIV infection occurs frequently have a static encephalopathy for a number of reasons unrelated to viral infection, the most common being the complications of premature birth. The category of normal neurological examination is self-explanatory. Some children who initially were classified as having a static encephalopathy or a normal neurological examination have developed progressive encephalopathy while under surveillance.

In general the degree of neurological involvement tends to parallel the severity of the immunodeficiency. The majority of children with progressive encephalopathy have AIDS[7]. It is important to point out however that some children developed progressive neurological signs while diagnosed as having ARC and infrequently children present with progressive encephalopathy while immunologically asymptomatic but seropositive for HIV[7]. The onset of progressive encephalopathy in children with HIV infection indicates a poor prognosis and predicts a fatal outcome[7].

NEUROPATHOLOGICAL OBSERVATIONS

A unique constellation of neuropathological findings has been described including grossly diminished brain weight, inflammatory cell infiltrates with multinucleated giant cells and diffuse white matter astrocytosis. In addition, vascular calcification and inflammation primarily in the basal ganglia and periventricular white matter are seen corresponding to the calcified lesions noted on CT scans. Opportunistic or reactivated latent infections were found infrequently.

The multinucleated cells (Figure 5.1A) are the hallmark of HIV infection of brain and are thought to be derived from invading macrophages which have undergone syncytial formation[11, 24]. They have been shown by electron microscopy to contain mature virions as well as immature and rare budding forms[25, 26] (Figure 5.1C). Subsequently these macrophages and multinucleated giant cells have been shown to harbour HIV genome[27, 28] (Figure 5.1B) and to express HIV antigen[28, 29].

The only other cell types which have been shown to contain HIV genome or antigen are endothelial and perithelial cells of cerebral blood vessels[9, 27]. Preliminary data suggests that a very small number of neurons or glial cells may express HIV antigen but these findings require further confirmation[9, 27, 28].

HIV-SPECIFIC ANTIBODIES AND ANTIGEN IN THE CSF

In search of prognostic factors, we investigated intra-blood–brain barrier (IBBB) synthesis of antibodies to HIV and the presence of HIV-specific antigens in the CSF of children with HIV infection[30]. A summary of the data from this study is shown in Table 5.1.

IBBB synthesis of HIV antibodies was identified in children in all three designated neurological categories. This finding is interpreted as indicating viral invasion but did not show a significant correlation with the degree of neurological dysfunction. It is not known whether some of the children with IBBB synthesis of HIV antibodies who are currently neurologically normal, or who have a static encephalopathy, are destined to show neurological deterioration.

A strong correlation was observed between HIV antigen expression in the CSF and progressive encephalopathy. Eight of 11 children with progressive encephalopathy had HIV antigen in their CSF, while none of the children with static encephalopathy or a normal neurological examination had detectable antigen in their CSF.

Prospective longitudinal studies of HIV-specific antibodies and antigens in the CSF are still necessary to assess the prognostic value of these studies.

PATHOGENETIC MECHANISMS

Any theory of the pathogenesis of HIV brain infection in children must take into account the following clinical and neuropathological observations:

(1)   early brain infection and possible 'neurotropism';

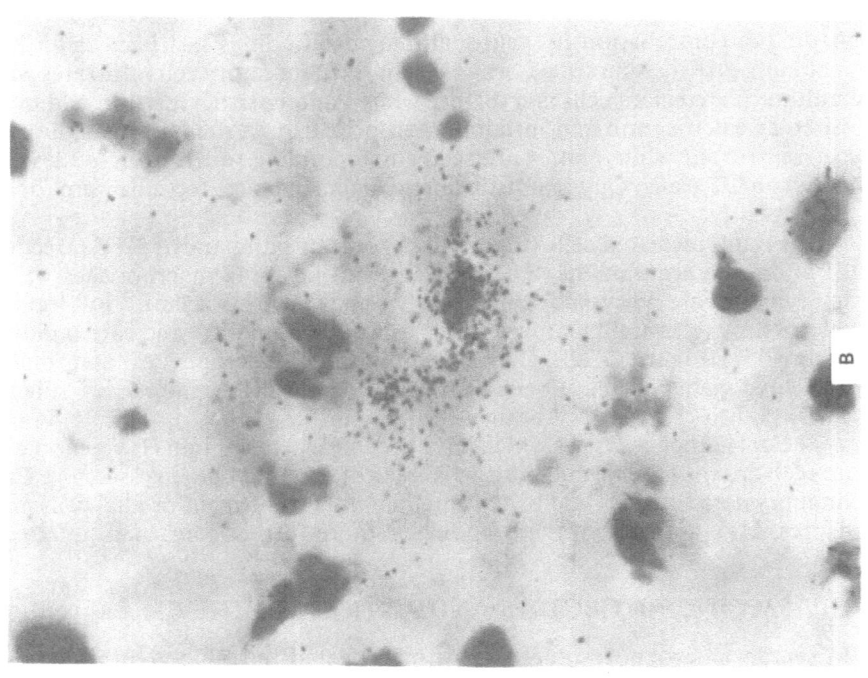

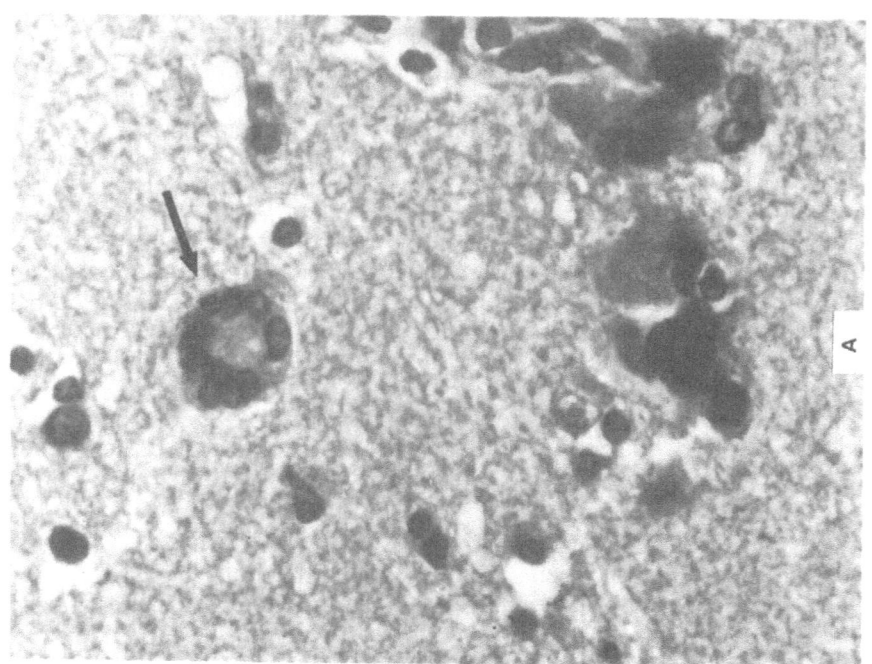

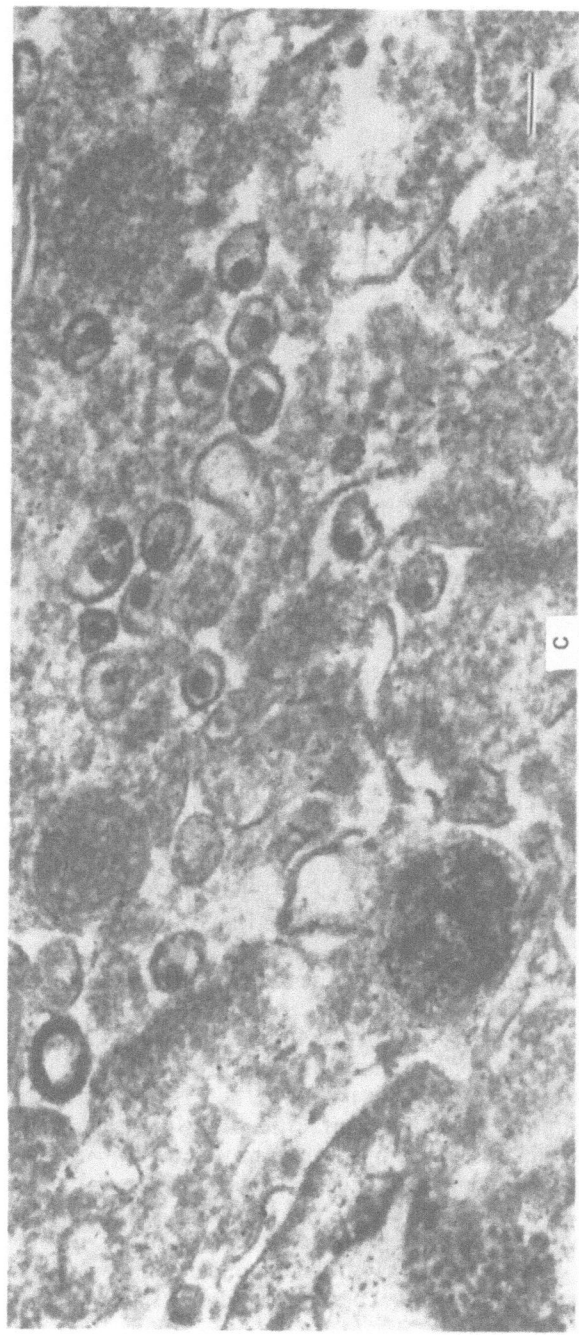

Figure 5.1 Post mortem brain tissue from a child who died with HIV infection and progressive encephalopathy: (A) a multinucleated giant cell (arrow) in basal ganglia, basophilic perivascular calcifications are also seen (H & E, ×800); (B) *in situ* hybridization using HIV (HTLV-III) RNA probe showing viral genome in an unidentified mononucleated cell (×800); (C) electron micrograph exhibiting immature and mature HIV virions in a multinucleated giant cell (bar = 100 nm; ×80000)

Table 5.1 HIV specific antibodies and antigen in serum and CSF of 27 children

Neurological status	Serum Ab	Serum Ag	CSF Ab[a]	CSF Ag
Progressive encephalopathy (n = 11)	10 (91%)	10 (91%)	7 (64%)	8 (73%)
Static encephalopathy (n = 9)	9 (100%)	6 (67%)	4 (44%)	0 (0%)
Neurologically normal (n = 7)	7 (100%)	2 (29%)	2 (29%)	0 (0%)

Ab = HIV antibody; Ag = HIV antigen
[a] Intra-blood–brain barrier synthesis
Data obtained in collaboration with J. Goudsmit, University of Amsterdam, and D. A. Paul, Abbott Laboratories

(2) the often long interval from initial infection to the onset of neurological dysfunction;

(3) the progressive but *stepwise* neurological deterioration;

(4) the apparent correlation of viral antigen expression in the CSF with progressive encephalopathy;

(5) the invasion of the brain by inflammatory cells, in particular macrophages, harbouring replicating retrovirus;

(6) the formation of syncytial multinucleated giant cells; and

(7) the scarcity of detectable antigen in cells of neuroectodermal origin.

A discussion of possible pathogenic mechanisms is best approached by posing certain questions.

How does HIV enter the brain?

Recent evidence suggests that HIV invades the brain early, probably at the time of initial infection in most if not all individuals[15, 31, 32]. Preliminary studies have shown that there is a peak of HIV antigen detected in the CSF soon after primary infection, and that infectious virus can be isolated at this time[15, 32].

How HIV enters the brain at this time may be difficult to determine from the study of post mortem brain tissue. Animal lentiviruses such as visna or caprine arthritis encephalomyelitis virus may initially infect choroid plexus cells in the former, and synovial lining cells, in the latter[17], however virions or viral antigens may not be easily found in these cells later in the course of the infection. Similarly, initial brain infection with HIV may occur in endothelial cells[9, 27], choroid plexus or possibly ependymal cells, followed by a slow dissemination of infection from these sites as seen in visna infection of sheep[17].

The only mechanism by which HIV may enter the brain for which substantial evidence exists is via HIV-infected macrophages[11, 25–29]. It is not known at this time if these cells enter the brain early, late, or episodically during the course of infection.

Is HIV neurotropic?

If one means by this simply 'Can HIV enter the brain?', the answer is clearly 'Yes'[11-16,25-29]. Using the more precise definition of neurotropic to indicate infection of cells of neuroectodermal origin (neurons or glia) this has not been convincingly demonstrated to date. Preliminary studies suggest that at least small numbers of glial cells and possibly neurons may contain viral antigen as determined by the immunoperoxidase technique[9,27], and neuropathologically we see damage to both grey and white matter structures suggesting that these tissues may be targets for immunocytopathic attack. This is not too dissimilar from visna where the presence of virus-infected cells (by *in situ* hybridization) corresponds with the magnitude of inflammatory cells infiltrates[17].

Which cells are infected?

Infection at the cellular level, in strict virological terms, is defined as attachment, penetration, and uncoating of viral genetic material. Thereafter, depending on the nature of the virus and the susceptibility of the cell, the virus may:

(1) actively replicate and lyse the cell;

(2) be expressed in a restricted fashion and persist within the cell; or

(3) enter a latent state where viral genome is present but viral antigens are not produced.

It appears that complete viral replication occurs in invading macrophages probably of monocyte lineage[11,24-29]. In cells of neuroectodermal origin, evidence from immunocytochemistry and *in situ* hybridization studies suggests that if viral infection does occur then the virus remains in a latent state, or that expression is severely restricted in these cells.

How is viral latency or expression controlled?

The answer to this question is still unknown but it is fundamental to the understanding of HIV infection.

A model of viral latency and reactivation in cultured lymphocytes has recently been reported[33]. In these lymphocytes HIV genome is present as stable integrated proviral DNA which can be induced months later to produce infectious virus. Control of viral expression at the transcriptional level has been suggested but the mechanism remains to be elucidated.

An alternative model for restricting viral expression comes from recent work demonstrating that visna-infected monocytes signal lymphocytes in close proximity to release a new type of interferon[34]. The action of this interferon is to restrict cell maturation, viral production and cell fusion. If a similar mechanism were operating in HIV-infected monocytes, as some recent evidence suggests[35], one might speculate that once these infected monocytes (macrophages) become situated in the neuropil, where they are free of the close proximity of lymphocytes, active replication of HIV would resume, resulting in cell fusion, syncytial giant cell formation, and cell lysis with further dissemination of infection.

What pathogenetic mechanism(s) can we postulate based on animal lentivirus models?

As we postulate pathogenetic mechanisms we need to keep in mind the clinical features described above, particularly the slowly progressive and stepwise neurological deterioration, the correlation of progressive encephalopathy with viral antigen expression in the CSF, and the neuropathological findings of invading HIV-infected macrophages, and cytopathological changes with syncytial giant cell formation in deep grey and white matter structures.

We must also look to the extensive literature on pathogenetic mechanisms of the animal lentiviruses which share features including: evasion of immune surveillance by remaining latent in certain cells as integrated proviral DNA, episodic reactivation, or restricted expression of viral antigens, associated with an immunologically mediated inflammatory response directed against these viral antigens, resulting in cumulative tissue damage[36].

Lastly, we must take into account the unique feature of HIV infection which is different from other lentiviruses, specifically with respect to the immunodeficiency present in the host.

A theoretical scenario to incorporate these features is as follows:

(1) The brain is infected early, perhaps at the time of initial HIV infection, possibly involving endothelial cells, choroid plexus or ependymal cells. (This may be related to the role of certain of these cells in Ia presentation and antigen recognition.) Initial infection may be clinically silent or result in meningoencephalitis[15,31,32]. During this phase an initial peak in antigen occurs[32] and virus can be isolated from the CSF[15].

(2) This is followed by a rise in HIV antibodies titres within the CSF[32] and probably some clearing of HIV antigen and free virus.

(3) HIV may then persist in a latent form as integrated proviral DNA or with restricted viral expression, possibly in neurons or glial cells, which are particularly stable and suitable for this type of infection.

(4) Episodic reactivation of viral expression and the production of viral antigens would lead to the entry of macrophages into the brain, in their role as effector cells in the immunological defence of the host. These macrophages, however, are a Trojan horse, or perhaps are better designated 'double agents' since these are virus-infected cells. The macrophages contribute directly to the cytopathic process by known effector mechanisms but in addition form syncytial giant cells, resulting in further local dissemination of HIV infection to surrounding cells.

(5) Cycles of latency or restricted expression may alternate with those of inflammation, each amplified by the entry of viral-infected macrophages, while at the same time attenuated by the defective cell-mediated immune response seen in the later stages of HIV infection.

(6) The distribution of the lesions seen neuropathologically may be explained by a slow ventriculo-fugal spread of infection, perhaps similar to that seen with visna, which would account for the damage to deep grey and white matter structures.

In summary, then, as with other lentiviruses, the failure of the host to remove non-expressing provirus-infected cells would allow for subsequent expression of viral antigens and episodic bouts of inflammation resulting in cumulative tissue damage.

Acknowledgements

Peter Dowling and Joseph Menonna performed the *in situ* hybridization studies illustrated in Figure 5.1. Markus Meyenhofer assisted with the electron micrograph. Leroy Sharer is thanked for review of the manuscript and critical comments. This work was supported, in part, by PHS grant # AI 23242-01.

References

1. Barre-Sinoussi, F., Chermann, J. C., Rey, F., Nugeyre, M., Chamaret, S., Gruest, J., Dauquet, C., Axler-Blin, C., Vezinet-Brun, F., Rouzioux, C., Rozenbaum. W. and Montagnier, L. (1983). Isolation of a T-lymphotropic retrovirus from a patient at risk for acquired immune deficiency syndrome (AIDS). *Science*, **220**, 868–71
2. Popovic, M., Sarngadharan, M. G., Read, E. and Gallo, R. C. (1984). Detection, isolation, and continuous production of cytopathic retroviruses (HTLV-III) from patients with AIDS and pre-AIDS. *Science*, **224**, 497–500
3. Levy, J. A., Hoffman, A. D., Kramer, S. M., Landis, J. A., Shimabukuro, J. M. and Oshiro, L. S. (1984). Isolation of lymphocytopathic retroviruses from San Francisco patients with AIDS. *Science*, **225**, 840–84
4. Coffin, J., Hasse, A., Levy, J. A., Montagnier, L., Oroszlan, S., Teich, N., Temin, H., Toyoshima, K., Varmus, H., Vogt, P. and Weiss, R. (1966). Human immunodeficiency viruses. (Letter). *Science*, **232**, 697
5. Epstein, L. G., Sharer, L. R., Joshi, V. V., Fojas, M., Koenigsberger, M. R. and Oleske, J. M. (1985). Progressive encephalopathy in children with acquired immune deficiency syndrome. *Ann. Neurol.*, **17**, 488–96
6. Belman, A. L., Ultmann, M. H., Horoupian, D., Novick, B. E., Spiro, A. J., Rubinstein, A., Kurtzberg, D. and Cone-Wesson, B. (1985). Neurological complications in infants and children with acquired immune deficiency syndrome. *Ann. Neurol.*, **18**, 560–6
7. Epstein, L. G., Sharer, L. R., Oleske, J. M., Connor, E. M., Goudsmit, J., Bagdon, L., Robert-Guroff, M. and Koenigsberger, M. R. (1986). Neurologic manifestations of HIV infection in children. *Pediatrics*, **78**, 678–87
8. Navia, B. A., Jordan, B. D. and Price, R. W. (1986). The AIDS dementia complex: I. Clinical features. *Ann. Neurol.*, **19**, 517–24
9. Navia, B. A., Cho, E.-S., Petito, C. K. and Price, R. W. (1986). The AIDS dementia complex: II. Neuropathology. *Ann. Neurol.*, **19**, 525–35
10. Epstein, L. G., Sharer, L. R. and Gajdusek, D. C. (1986). Hypothesis: AIDS encephalopathy is due to primary and persistent infection of the brain with a human retrovirus of the lentivirus subfamily. *Med. Hypoth.*, **21**, 87–96

11. Sharer, L. R., Epstein, L. G., Cho, E.-S., Joshi, V. V., Meyenhofer, M. F., Rankin, L. F. and Petito, C. K. (1986). Pathologic features of AIDS encephalopathy in children: Evidence for LAV/HTLV-III infection of brain. *Hum. Pathol.*, **17**, 271-84

12. Shaw, G. M., Harper, M. E., Hahn, B. H., Epstein, L. G., Gajdusek, D. C., Price, R. W., Navia, B. A., Petito, C. K., O'Hara, C. J., Groopman, J. E., Cho, E.-S., Oleske, J. M., Wong-Staal, F. and Gallo, R. C. (1985). HTLV-III infection in brains of children and adults with AIDS encephalopathy. *Science*, **227**, 177-82

13. Gajdusek, D. C., Amyx, H. L., Gibbs, C. J., Asher, D. M., Rodgers-Johnson, P., Epstein, L. G., Sarin, P. S., Gallo, R. C., Maluish, A., Arthur, L. O., Montagnier, L. and Mildvan, D. (1985). Infection of chimpanzees by human T-lymphotropic retroviruses in brain and other tissues from AIDS patients. *Lancet*, **1**, 55-6

14. Levy, J. A., Shimabukuro, J., Hollander, H., Mills, J. and Kaminsky, L. (1985). Isolation of AIDS associated retrovirus from cerebrospinal fluid and brain of patients with neurological symptoms. *Lancet*, **2**, 586-8

15. Ho, D. D., Rota, T. R., Schooley, R. T., Kaplan, J. C., Allan, J. D., Groopman, J. E., Resnick, L., Felsenstein, D., Andrews, C. A. and Hirsch, M. S. (1985). Isolation of HTLV-III from CSF and neural tissues of patients with AIDS related neurologic syndromes. *N. Engl. J. Med.*, **313**, 1493-7

16. Salahuddin, Z. S., Markham, P. D., Popovic, M., Sarngadharan, M. G., Orndorff, S., Fladagar, A., Patel, A., Gold, J. and Gallo, R. C. (1985). Isolation of infectious human T-cell leukemia/lymphotropic virus type III (HTLV-III) from patients with acquired immunodeficiency syndrome (AIDS) or AIDS-related complex (ARC) and from healthy carriers: A study of risk groups and tissue sources. *Proc. Natl. Acad. Sci. USA*, **82**, 5530-4

17. Narayan, O. and Cork, L. C. (1985). Lentiviral diseases of sheep and goats: chronic pneumonia, leukoencephalomyelitis, and arthritis. *Rev. Infect. Dis.*, **7**, 89-98

18. Letvin, N. L., Daniel, M. D., Sehgal, P. K., Desrosiers, R. C., Hunt, R. D., Waldron, L. M., MacKey, J. J., Schmidt, D. K., Chalifoux, L. V. and King, N. W. (1985). Induction of AIDS-like disease in macaque monkeys with T-cell tropic retrovirus STLV-III. *Science*, **230**, 71-3

19. Scott, G. B., Buck, B. E., Leterman, J. G., Bloom, F. L. and Parks, W. P. (1984). Acquired immunodeficiency in infants. *N. Engl. J. Med.*, **310**, 76-82

20. Jovaissas, E., Koch, M. A., Schafer, A., Stauber, M. and Lowenthal, D. (1985). LAV/HTLV-III in 20-week fetus. *Lancet*, **2**, 1129

21. Marion, R. W., Wiznia, A. A., Hutcheon, R. G. and Rubinstein, A. (1986). Human T-cell lymphotropic virus type III (HTLV-III) embryopathy. *Am. J. Dis. Child*, **140**, 638-40

22. Belman, A. L., Lantos, G., Horoupian, D., Novick, B. E., Ultmann, M. H., Dickson, D. W. and Rubinstein, A. (1986). AIDS: Calcification of the basal ganglia in infants and children. *Neurology*, **36**, 1192-9

23. Epstein, L. G., Sharer, L. R., Berman, C. Z. and Khademi, M. (1987). Unilateral calcification and contrast enhancement of the basal ganglia in a child with AIDS encephalopathy. *Am. J. Neuroradiol.*, **8**, 163-5

24. Sharer, L. R., Cho, E.-S. and Epstein, L. G. (1985). Multinucleated giant cells and HTLV-III in AIDS encephalopathy. *Hum. Pathol.*, **16**, 760

25. Epstein, L. G., Sharer, L. R., Cho, E.-S., Meyenhofer, M. F., Navia, B. A. and Price, R. W. (1985). HTLV-III/LAV-like retrovirus particles in the brain of patients with AIDS encephalopathy. *AIDS Res.*, **1**, 447-54

26. Meyenhofer, M. F., Epstein, L. G., Cho, E.-S. and Sharer, L. R. (1987). Ultrastructural morphology and intracellular production of HIV in brain. *J. Neuropathol. Exp. Neurol,* **46**, 474-84

27. Wiley, C. A., Schrier, R. D., Nelson, J. A., Lampert, P. W. and Oldstone, M. B. A. (1986). Cellular localization of human immunodeficiency virus infection in the brains of acquired immune deficiency syndrome patients. *Proc. Natl. Acad. Sci. USA*, **83**, 7089-93

28. Koenig, S., Gendelman, H. E., Orenstein, J. M., Dal Canto, M. C., Pezeshkpour, G. H., Yungbluth, M., Janotta, F., Aksamit, A., Martin, M. A. and Fauci, A. S. (1986). Detection of AIDS virus in macrophages in brain tissue from AIDS patients with encephalopathy. *Science*, **233**, 1089-93

29. Gabuzda, D. A., Ho, D. D., de la Monte, S. M., Hirsh, M. S., Rota, T. R., Sobel, R. A. (1986). Immunohistochemical identification of HTLV-III antigen in brains of patients with AIDS. *Ann. Neurol.*, **20**, 289-95

30. Epstein, L. G., Goudsmit, J., Paul, D. A., Morrison, S., Connors, E. M., Oleske, J. M. and Holland, B. (1987). HIV expression in the cerebrospinal fluid of children with progressive encephalopathy. *Ann. Neurol.*, **21**, 397–401
31. Goudsmit, J., Wolters, E. C., Bakker, M., Smit, L., van der Noordaa, J., Hische, E. H., Tutuarima, J. and van der Helm, H. J. (1986). Intrathecal synthesis of antibodies to HTLV-III in patients without AIDS or AIDS related complex. *Br. Med. J.*, **292**, 1231–4
32. Goudsmit, J., De Wolf, F., Paul, D. A., Epstein, L. G., Lange, J. M. A., Krone, W. J. A., Speelman, H., Wolters, E. C., Van Der Noordaa, J., Oleske, J. M., Van Der Helm, H. J. and Coutino, R. A. (1986). Expression of human immunodeficiency virus antigen (HIV-Ag) in serum and cerebrospinal fluid during acute and chronic infection. *Lancet*, **2**, 177–80
33. Folks, T., Powell, D. M., Lightfoote, M. M., Benn, S., Martin, M. M. and Fauci, A. S. (1986). Induction of HTLV-III/LAV from a nonvirus producing T-cell line: Implications for latency. *Science*, **231**, 600–2
34. Narayan, O. (1986). Novel mechanism for induction of interferon by visna lentiviruses: possible role of the interferon in restricted "slow" replication and chronic disease. In Salzman, L. A. (ed.) *Animal Models of retrovirus Infection and Their Relationship to AIDS*. pp. 355–66. (Orlando: Academic Press)
35. Eyster, M. E., Goedert, J. J., Poon, M. C. and Preble, O. T. (1983). Acid-labile α-interferon. A possible preclinical marker for the acquired immunodeficiency syndrome in hemophilia. *N. Engl. J. Med.*, **309**, 583–6
36. Clements, J. E. (1985). Hypothesis on the molecular basis of nononcogenic retroviral diseases. *Rev. Infect. Dis.*, **7**, 68–74

6
Persistent Infection of Mice with Lymphocytic Choriomeningitis Virus

F. LEHMANN-GRUBE

INTRODUCTION

When speaking of virus diseases, one usually thinks of the consequences of acute infections ending either fatally or with recovery, in which case the virus is usually rapidly eliminated. Obviously acute virus infections and their pathological effects do occur frequently, and in many epidemics of infectious diseases of animals and man viruses were and still are aetiologically responsible. There are, however, other viruses that have the ability to persist, often for life. Their pathogenetic potentials are not as obvious, although many illnesses are known to be caused by chronic infections with such agents; in others the viral aetiology is suspected, and it could well be that persisting viruses are more relevant for ailments of animals and man than has hitherto been assumed. Thus, prevention and therapy of chronic virus diseases may turn out to be major tasks for physicians as well as virologists; both are poorly equipped for this. Our understanding of the mechanisms by which higher organisms control acute virus infections is limited and we know even less of the other side of the coin, namely how certain viruses evade the host's defences. In the case of illnesses caused by the so-called unconventional viruses ignorance precludes the formulation of a hypothesis that would direct rational attempts at prophylaxis, not to mention a cure, once clinical signs have developed. With respect to slow virus diseases caused by conventional viruses the prospects may not be as dismal, although here too great gaps in our knowledge exist. Thus, a considerable amount of work has yet to be done before these diseases are sufficiently understood for their control to become possible, and every model that promises to advance our understanding in this field is worth being exploited.

THE SYSTEM

The lymphocytic choriomeningitis (LCM) virus is the type species of the arena-viruses, a group of agents that are morphologically and biochemically similar and share many biological properties. The name (initially arenoviruses)

69

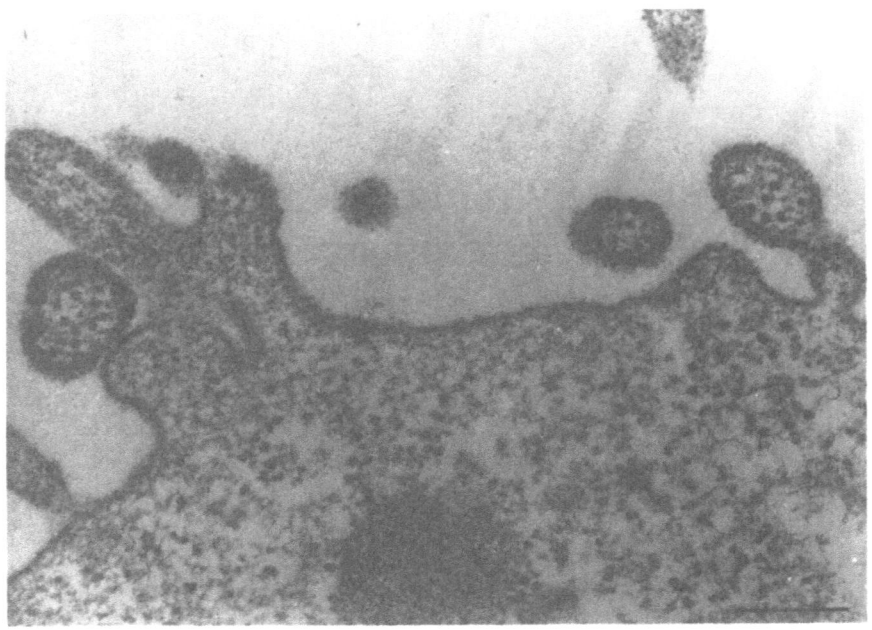

Figure 6.1 Electron micrograph of LCM virus-infected cultivated mouse L cell. Thirty-six hours after infection with strain WE LCM virus, cells were fixed with paraform–glutaraldehyde and osmium tetroxide solutions, dehydrated and embedded in ERL-resin. Ultrathin section was contrasted with lead citrate and uranyl acetate; scale bar represents 0.2 μm. Kindly supplied by Dr Klaus Mannweiler, Hamburg

reflects the characteristic granules seen by electron microscopy in sections of infected cells (Figure 6.1). The approved family name is Arenaviridae, with only one genus at present, Arenavirus[1]. At the time of writing, the genus comprised 13 members belonging to two 'complexes' called LCM-Lassa (LCM, Lassa, Mopeia and Mobala viruses) and Tacaribe (Amapari, Junin, Latino, Machupo, Parana, Pichinde, Tacaribe, Tamiami and Flexal viruses) (W. R. Rawls, personal communication). Of these, LCM virus, Junin virus, Machupo virus and Lassa virus are pathogenic for man causing, respectively, lymphocytic choriomeningitis, Argentine haemorrhagic fever, Bolivian haemorrhagic fever, and Lassa fever[2].

Many of the arenaviruses have proved to be of great value for studying a variety of phenomena of relevance in virology, immunology and medicine. With different experimental animals numerous virus–host combinations have been examined. While all have added to our knowledge most has been learned from the first and best analysed example, the LCM virus-infected *Mus musculus*[3].

If an adult mouse is infected for the first time with the LCM virus, usually illness and often death ensue; the surviving animal clears the agent with the aid of its immune system and is subsequently protected. In contrast, if the first encounter occurs before or immediately after birth a carrier state results

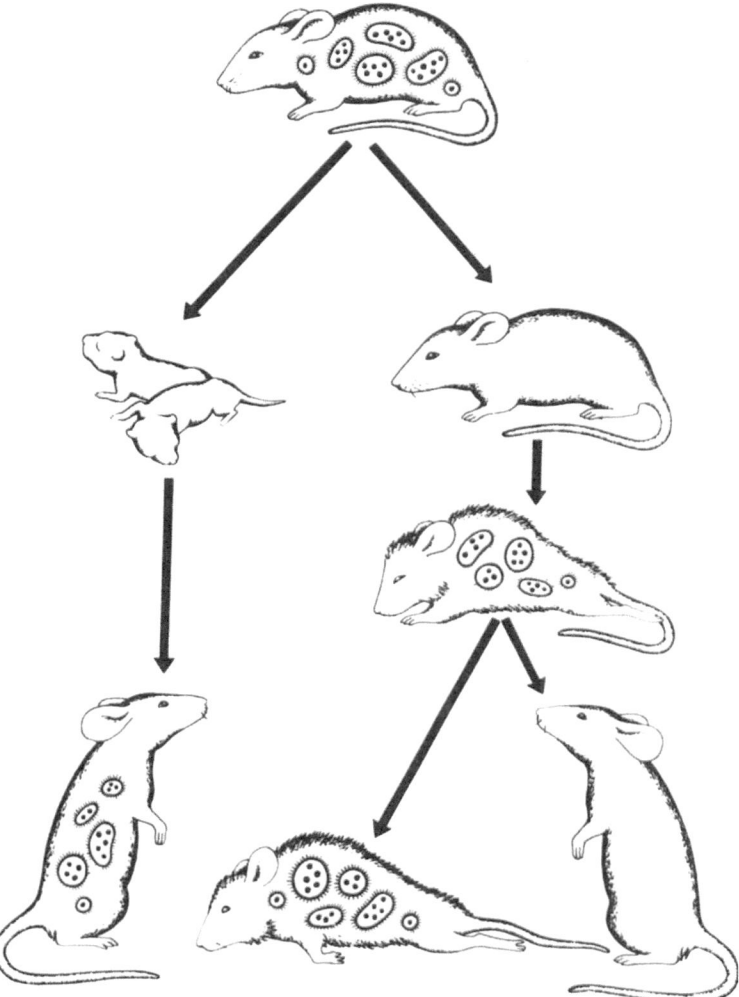

Figure 6.2 Principal phenomena seen in LCM virus-infected mice. From Lehmann–Grube[4] with permission of Academic Press, New York

in which large quantities of virus are produced and persist lifelong in all organs (Figure 6.2)[4,5]. In ageing carrier mice of certain (but not all) strains, a chronic 'late-onset disease' may develop[6].

These phenomena, which on first sight are paradoxical, are explained as follows. The virus is not directly harmful for the cells in which it multiplies, and illness and death of acutely infected adult mice are expressions of excessive immune reactions against viral or virus-induced antigens[7]. Persistence of the virus and absence of illness, on the other hand, result from LCM virus-specific immunological tolerance that develops if the animal is immunologically immature when making its first contact with the agent[8].

The tolerance is restricted to the T cell compartment, because antiviral antibodies are produced and are responsible for the late-onset disease, which, actually, is an immune complex disease[9].

The LCM virus carrier mouse appears uniquely suited for studying the mechanism that allows a virus and its host to coexist: it is a natural phenomenon *par excellence* which can easily be reproduced in the laboratory; mice are bred without difficulty and numerous genetically defined strains exist allowing experimental analysis to a degree not possible with any other mammalian host; and the virus has been well studied and its structure is sufficiently known to allow analysis of the interaction between viral components and the host's immune system[10-12].

A virus can persist for prolonged periods of time provided:

(1) its replication is so regulated that uninhibited expansion is prevented but replenishment of losses due to natural decay and elimination are permitted;

(2) vital tissues remain functionally intact, at least to a degree that is compatible with the host's survival;

(3) the immune system of the organism is hindered from fulfilling its function, i.e. eliminating the agent[13].

Of these three requirements, the role of the immune system will be further discussed here.

LCM VIRUS-SPECIFIC IMMUNOLOGICAL TOLERANCE IN LCM VIRUS CARRIER MICE

Elimination of LCM virus from acutely infected adult mice is mediated by cytotoxic T-lymphocytes (CTL); antibodies against the virus are produced but play little or no role in the clearance mechanism[14]. LCM virus carriers cannot get rid of the virus because LCM virus-specific CTL activities are absent (more correctly, cannot be detected) in these animals[5], although antibodies directed against the virus have often been demonstrated[15-17]. Thus, the immunological unresponsiveness of carrier mice to LCM virus is restricted to a subpopulation within the T cell compartment and hence is an example of split tolerance[18] or immune deviation[19] – but it should be stressed that the relationship between the immunogens activating LCM virus-specific CTL precursors and the immunogens inducing antibodies (probably by activating helper cells) has not yet been defined. The immunological unresponsiveness of carrier mice is specific, because these animals react quite normally to most other immunogens[5].

Lifelong persistence of LCM virus in neonatal or congenital carrier mice (and erythrocyte chimerism of cattle twins[20]) were the phenomena on which Burnet and Fenner based their concept of the development of self-recognition during ontogeny[21], later to merge with Medawar's experimentally induced 'actively acquired tolerance'[22]. Immunological tolerance is an operational term, and several mechanisms have been described. Specific suppressor cells or blocking factors were not found in LCM virus carrier mice, which was considered evidence for clonal inactivation, possibly resulting from the mouse's acceptance of the virus as self[23], although the mechanism remained speculative. *A priori*, one would not expect a virus to be a tolerogen, and indeed few such agents are known to induce immunological tolerance in the mouse or in members of other species. The LCM virus, too, is markedly immunogenic in the adult mouse and only causes the development of tolerance in the immunologically immature or compromised animal. Nonetheless, it is usually assumed that the absence in carrier mice of cell-mediated immunity directed against the LCM virus is not principally different from numerous other examples of naturally occurring or experimentally induced immunological tolerance where the tolerogens are not infectious. The finding, by Popescu *et al.*[24,25] and Doyle and Oldstone[26], that a few apparently resting lymphocytes in carrier mice are always infected suggested the possibility that the inability to eliminate the virus had a virological basis.

Since Traub's early description of carrier mice[27] it has been known that such animals are viraemic. It turned out that most of the viral infectivity in the blood was associated with lymphocytes, and a closer look revealed that a small proportion of T-lymphocytes in blood as well as lymphoid organs of carrier mice always harboured the virus. Infectivity could only be revealed by assaying intact viable cells; as soon as these were disrupted or inactivated by means known not to affect the virus, little if any cell-associated ineffectivity could be detected[25]. Both T- and B-lymphocytes were found to be infected, but in subsequent work we concentrated on the former because these are essential for virus elimination[28].

We never succeeded in infecting T-lymphocytes of adult mice either *in vitro* or *in vivo*, and attempts at demonstrating LCM virus receptors on these cells have consistently failed. In contrast, in unborn or newborn mice it was always possible to infect a few T-lymphocytes, and this infection persisted when the animals grew older (Figure 6.3). The T-lymphocytes' susceptibility decreased with increasing age of the mice, reaching zero levels when the animals were 3 weeks old (Figure 6.4), which correlates well with the proportion of mice becoming carriers upon experimental infection. Further studies[29] disclosed that immunological activation of T-lymphocytes in adult mice during a host-*versus*-graft reaction did not result in their infection and that the T-lymphocytes among the blast cells in the spleens of adoptively immunized cyclophosphamide-induced carrier mice also resisted infection. When newborn mice were infected, initially the majority of infectious T-lymphocytes were detected in the thymus, but their numbers declined and, when the animals were 5 weeks old and older, they were only irregularly found among thymocytes. In contrast, the numbers of infectious T-lymphocytes in the spleens remained constant. In 22 mice, sacrificed between 21 and 63 days after neonatal infection, an average of 320 T-lymphocytes in 10^6 splenocytes scored as infectious centres on L cells. This figure would have been four times higher had the titrations been performed in the mouse[25], and approximately 40% of nucleated cells in the mouse's spleen are T-lymphocytes[30]; from this, it can be estimated that roughly 0.05% of the spleen T-lymphocytes of established neonatal carrier mice are infected.

73

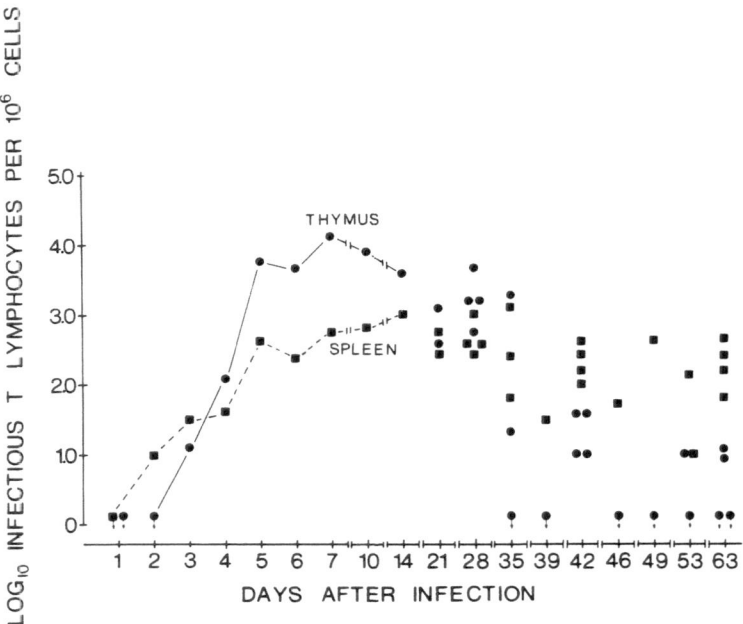

Figure 6.3 Appearance and persistence of infectious T-lymphocytes in thymi and spleens of carrier mice. NMRI mice at age less than 24 h were infected intraperitoneally with strain WE LCM virus. At intervals the proportion of infectious T-lymphocytes among nucleated cells was calculated by subtracting numbers of infectious centres remaining after treatment of cells with anti-Thy-1 antibody plus complement from numbers of infectious centres forming with untreated cells. For two weeks after infection, data points represent pooled cells from several mice; thereafter, results are for individual mice. From Lehmann-Grube et al.[29] with permission from The Society for General Microbiology, UK

Another difference between spleen and thymus is worth mentioning. The highest virus concentration in the former was $10^{9.5}$ plaque-forming units per g of homogenized tissue while the corresponding value in the thymus, although attained 3 days later, was $10^{10.2}$. Preferential replication of LCM virus in the thymus of neonatally infected mice was also observed by Popescu and Ostrow[31] who, furthermore, found that infected thymocytes could initially be agglutinated by peanut lectin, a property that was lost when the mice grew older.

These observations have led to the formulation of a hypothesis[5,29] in which it is assumed that most murine T-lymphocytes lack *viral* receptors for adsorption of the virus as the first step of infection, but that the few cells that are programmed to respond immunologically against components of the viral envelope (presumably[32] this is gp44) bind the LCM virus via their immunological receptors; the resulting activation allows infection with the consequence of functional inactivation. This virus–T-lymphocyte interaction being primarily of an immunological nature, we postulate clonal expansion of the immunologically triggered and at the same time infected lymphocytes.

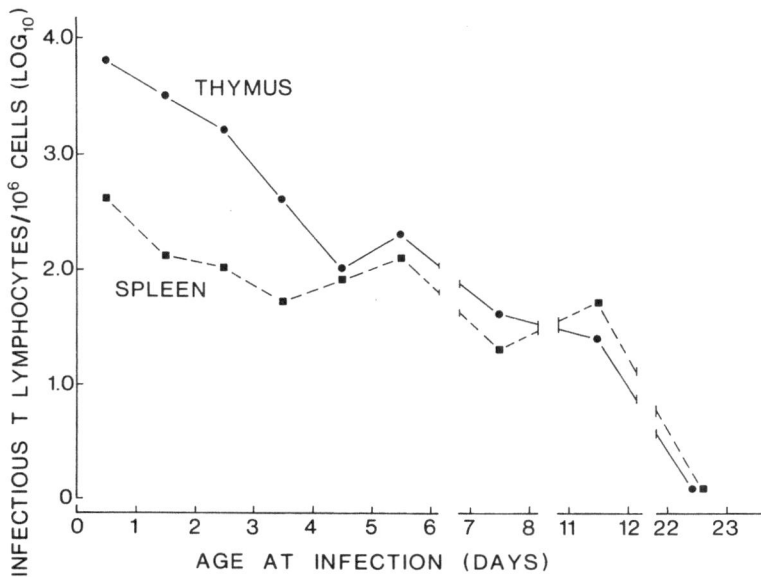

Figure 6.4 Effect of age of mice at infection with LCM virus on numbers of infectious T-lymphocytes in thymi and spleens. NMRI mice were infected by intraperitoneal inoculation with WE strain virus at intervals after birth. Five days later the proportion of infectious T-lymphocytes amongst the nucleated cells was determined essentially as has been described for Fig. 6.3. Modified from Tijerina *et al.*[28] with permission of Verlag der Zeitschrift für Naturforschung, Tübingen

We also assume that the virus is presented by cells, the normal function of which is antigen presentation (presumably mononuclear phagocytes, which are among the first cells undergoing infection in the mouse[33]). Taking into account the restriction of infectibility of T-lymphocytes to the time before and immediately after birth, the initial agglutinability by peanut lectin, and the preferential replication in the thymus, it is assumed that susceptibility of the T-lymphocytes is confined to an intrathymic stage of development, during which antigen receptors are functioning but full maturity has not yet been attained.

While this hypothesis is at present not at all proved, I am not aware of findings that would be incompatible. Admittedly, it is still possible that the susceptibility among immature murine T-lymphocytes is randomly distributed or that it is a property of an early developmental stage of a subclass of the cells unrelated to their immunological specificity. There is one observation, however, which is not easily reconciled with either of these explanations[29]. Persistence of LCM virus can be achieved by infecting immunosuppressed adult mice[5]; the protocol of Cole *et al.*[34] has been found particularly useful in this regard. Adult mice are infected by intracerebral inoculation; treatment 3 days later with a single dose of cyclophosphamide not only protects them against the characteristic immunopathological central nervous system disease but also leads to persistent infection. Unlike the carrier state that develops in congenitally or neonatally infected mice, in an infected and cyclophosphamide-treated adult mouse a mature immune system is prevented from responding to the virus, in all probability by preferential destruction of immunologically activated anti-viral T-lymphocyte precursors[35]. According to the hypothesis, such drug-induced carriers should be free of infected T-lymphocytes, which has indeed been found to be the case; 4 weeks after infection and treatment, when the mice had functionally and morphologically recovered (except for being unable to eliminate the LCM virus), infectious T-lymphocytes were not found in either thymuses or spleens; later, they appeared in low and erratic numbers[29], which was not unexpected because T-lymphocyte precursors with specificity for LCM virus were likely to emerge from stem cells; in a highly infectious environment these were bound to meet infectious virus resulting in immunological activation, infection, and functional incapacitation.

ANTIVIRAL ANTIBODY-PRODUCING CELLS IN PARENCHYMATOUS ORGANS OF LCM VIRUS CARRIER MICE

As has already been pointed out, in ageing carrier mice of certain mouse strains a 'late-onset disease'[6] develops, which has more recently been identified as immune complex disease (ICD)[9]; the characteristics of this syndrome have been described[5]. In addition to the changes caused by deposition of antigen–antibody complexes, in carrier mice with ICD essentially all organs contain mostly nodular, often extensive mononuclear cell infiltrates harbouring many plasma cells which have been regarded as an expression of pathological cell-mediated immune reactions[36] or of some immunoproliferative disease process[37]. Perhaps these infiltrates are the basis for the sometimes expressed opinion that mice persistently infected with LCM virus are predisposed to develop lymphomas[38,39]. My colleagues Demetrius Moskophidis and Jürgen Löhler and I have analysed these cell accumulations histologically and immunocytologically and have come to the conclusion that they represent functionally active lymphatic tissue (Figure 6.5). This finding suggested the possibility that the lymphoid infiltrates in the parenchymatous organs of carrier mice might participate in the humoral antiviral immune response.

Using a solid-phase immunoenzymatic technique for the localization and enumeration of single cells producing specific antibodies[40], we have found cells forming LCM virus-specific antibodies (AFC) in brains, kidneys, livers and spleens, but almost never in the blood (thereby excluding the possibility that the AFC demonstrated in the organs were derived from the circulation), of neonatal carrier mice of 10 strains and of conatal carrier house mice. The further analysis revealed that:

(1) most AFC produced IgG, although there was a minority producing IgM, which is in line with the previous finding that immune complexes in the kidneys of LCM virus carrier mice may contain IgM[41];

(2) there were considerable quantitative differences between AFC in the

76

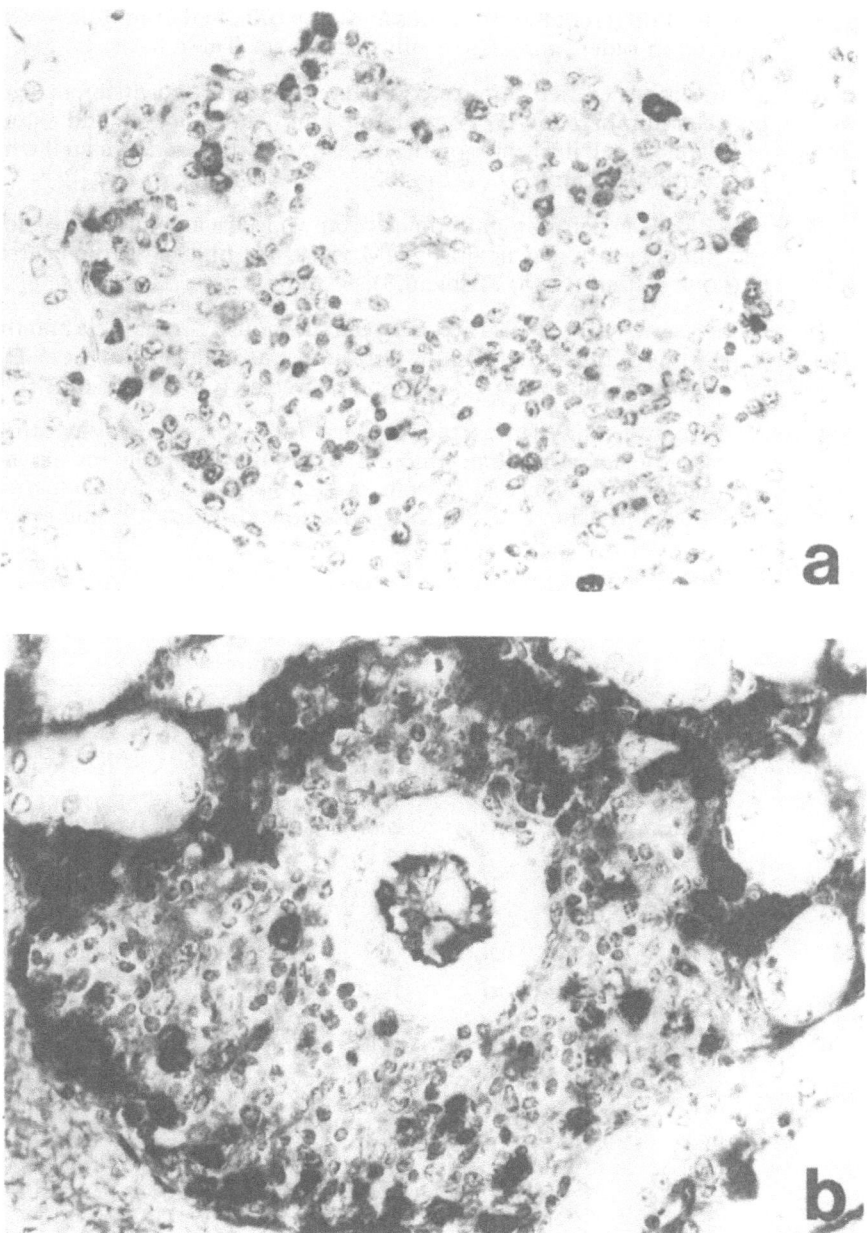

Figure 6.5 Immunoglobulin-producing cells in a kidney infiltrate of a 12-month-old neonatal LCM virus carrier mouse: (a) and (b) are sequential sections, stained for IgM and IgG, respectively. There are small numbers of IgM-positive cells which are localized predominantly at the margin of the infiltrate; IgG-producing cells are much more numerous. (PAP technique, counterstaining with a haemalum; ×360.) Kindly supplied by Dr Jürgen Löhler, Hamburg

organs of different mouse strains and, as a rule, higher numbers were counted in older animals (for illustration see Table 6.1);

(3) a correlation existed in mice of different strains and different ages between numbers of AFC on the one hand and numbers and extent of lymphoid cell infiltrates on the other and both were correlated with the severity of the ICD (Table 6.2);

(4) of the cells generating antibodies of any specificity, the proportion that made antibodies against LCM virus was highest in the central nervous system (CNS) (Table 6.3);

(5) there was no strict association between the mouse's haplotype and the magnitude of the antiviral immune response as judged from the extent of infiltrates, numbers of AFC, and degree of ICD (Table 6.2)[42,43].

We consider these findings important because they open the way to study experimentally the intrathecal production of immunoglobulin in experimental and natural CNS illnesses of man and animals[44,45]. This production has been observed in acute and subacute infectious diseases[46-50] and seems

Table 6.1 Numbers of cells producing IgG anti-LCM virus antibodies in organs of mice persistently infected with LCM virus[a]

Mouse strain	Organ	Age of mice (weeks)						
		7	19	26	31	36	42	58
NMRI	Brain	9[b]	273	308	564	619	646	643
	Kidney	<	193	125	278	80	307	1070
	Liver	8	45	21	235	99	140	90
	Spleen	136	148	378	655	363	560	687
	Blood	<	<	<	<	<	4	<
SWR	Brain	55	102	60	520	555	654	n.d.
	Kidney	60	264	100	138	320	323	n.d.
	Liver	10	103	20	25	<	143	n.d.
	Spleen	296	284	367	50	58	167	n.d.
	Blood	<	<	<	<	<	<	n.d.
C3H	Brain	<	<	<	<	55	10	n.d.
	Kidney	<	<	10	35	<	15	n.d.
	Liver	<	4	<	<	<	5	n.d.
	Spleen	<	62	30	95	25	105	n.d.
	Blood	<	<	<	<	<	<	n.d.
CBA	Brain	<	<	<	<	10	<	<
	Kidney	<	<	<	109	5	20	<
	Liver	<	<	<	61	40	10	<
	Spleen	8	<	<	37	<	35	<
	Blood	<	<	<	<	<	<	<

[a] Neonatal carrier mice established by intraperitoneal infection with WE strain LCM virus within 24 h after birth
[b] Means of numbers of antibody-forming cells per 10^6 trypan blue-excluding leukocytes in three to six mice
< = below detectability; n.d. = not determined

Table 6.2 Numbers of cells forming antibodies against LCM virus (AFC), numbers and sizes of infiltrates, and severity of immune complex disease (ICD) in parenchymatous organs of LCM virus carrier mice[a]

Mouse strain	H-2[b]	AFC	Infiltrates	ICD
CBA/J	k	+	+	+
C3H/HeJ	k	+	+	+
AKR/J	k	+	+	+
C57BR/cdJ	k	+ +	+ +	+ +
B10.BR/SgSnJ	k	+ +	+ +	+ +
NMRI	q	+ +	+ +	+ +
SWR/O1a	q	+ +	+ +	+ +
DBA/1LacJ	q	+ +	+ +	+ +
B10.G/O1a	q	+ +	+ +	+ +
C57BL/10SnJ	b	+ +	+ +	+ +
House mouse	–	(+)	+	–

[a] Carrier house mice were from a colony established 9 years ago with organ homogenate from a persistently infected wild *Mus musculus*; all other carriers were neonatally derived by inoculating WE strain LCM virus intraperitoneally within 24 h after birth
[b] Haplotype designation of the major histocompatibility gene complex

Table 6.3 Ratios of numbers of cells producing IgM, IgG and IgA antibodies against LCM virus to numbers of cells producing antibodies of the same classes with any specificity in organs of mice persistently infected with LCM virus[a]

Mice	Organ	Antibody-producing cells		
		LCM-specific	Total	Ratio
NMRI	Brain	603[b]	873	0.69
	Kidney	459	2.242	0.21
	Liver	190	820	0.23
	Spleen	533	14.617	0.04
	Blood	<	133	–

[a] Neonatal carrier mice at age 42 weeks established by intraperitoneal infection within 24 h after birth
[b] Means of numbers per 10^6 trypan blue-excluding leukocytes in three mice
< = below detectability

to be a regular finding when the course is chronic. In multiple sclerosis, antibodies with a variety of specificities have been detected[51,52], while during illnesses with known aetiologies antibodies against the causative agents are predominantly formed[53,54]. In certain slow virus diseases, such as subacute sclerosing panencephalitis[55,56], progressive rubella panencephalitis[57,58], and visna of sheep[59,60], antibodies directed against measles, rubella, and visna viruses, respectively, have been shown to be produced intracranially.

It is well established that antibody production requires the co-operation of several cell types; hence, if antibody is formed in the CNS these cells not only must have migrated there but also assembled locally. In cases of multiple sclerosis, the CNS has been reported to contain tissue resembling the immunoglobulin-secreting regions of lymph nodes[61], but nothing seems to be known of similar arrangements of cells of the immune system in brain diseases caused by persistent virus infections. There also seems to be no explanation, as yet, as to how these cells find their way through the blood-brain barrier into the CNS, nor is it easy to interpret the apparent longevity of certain B cell clones under these conditions[62,63].

We consider the LCM virus carrier mouse a valuable model for studying the mechanism that leads to the heterotopic production of antiviral antibodies during persistent virus infections.

Acknowledgements

Work done by the author and his colleagues was aided by research grants from the Deutsche Forschungsgemeinschaft and the Gemeinnützige Hertie-Stiftung zur Förderung von Wissenschaft, Erziehung, Volks- und Berufsbildung. The Heinrich-Pette-Institut is financially supported by Freie und Hansestadt Hamburg and Bundesministerium für Jugend, Familie, Frauen und Gesundheit.

References

1. Fenner, F. (1976). Classification and nomenclature of viruses. Second report of the International Committee on Taxonomy of Viruses. *Intervirology*, **7**, 1–115
2. Lehmann-Grube, F. (1988). Diseases of the nervous system caused by lymphocytic choriomeningitis virus and other arenaviruses. In Vinken, P. J., Bruyn, G. W. and Klawans, H. L. (eds.) *Handbook of Clinical Neurology, Viral Diseases.* (Amsterdam: Elsevier Science Publishers B.V.) (In press)
3. Lehmann-Grube, F. (1984). Portraits of viruses: arenaviruses. *Intervirology*, **22**, 121–45
4. Lehmann-Grube, F. (1982). Lymphocytic choriomeningitis virus. In Foster, H. L., Small, J. D., and Fox, J. G. (eds.) *The Mouse in Biomedical Research, Vol. II, Diseases.* pp. 231–66. (New York, London, Paris, San Diego, San Francisco, São Paulo, Sydney, Tokyo, Toronto: Academic Press)
5. Lehmann-Grube, F., Martínez Peralta, L., Bruns, M. and Löhler, J. (1983). Persistent infection of mice with the lymphocytic choriomeningitis virus. *Comp. Virol.*, **18**, 43–103
6. Hotchin, J. and Collins, D. N. (1964). Glomerulonephritis and late onset disease of mice following neonatal virus infection. *Nature*, **203**, 1357–9
7. Hotchin, J. (1962). The biology of lymphocytic choriomeningitis infection: virus-induced immune disease. *Cold Spring Harbor Symp. Quant. Biol.*, **27**, 479–99
8. Volkert, M. and Hannover Larsen, J. (1965). Immunological tolerance to viruses. *Progr. Med. Virol.*, **7**, 160–207
9. Oldstone, M. B. A. (1975). Virus neutralization and virus-induced immune complex disease. *Progr. Med. Virol.*, **19**, 84–119
10. Buchmeier, M. J., Welsh, R. M., Dutko, F. J. and Oldstone, M. B. A. (1980). The virology and immunobiology of lymphocytic choriomeningitis virus infection. *Adv. Immunol.*, **30**, 275–331
11. Bruns, M., Martínez Peralta, L. and Lehmann-Grube, F. (1983). Lymphocytic choriomeningitis virus. III. Structural proteins of the virion. *J. Gen. Virol.*, **64**, 599–611
12. Compans, R. W. and Bishop, D. H. L. (1985). Biochemistry of arenaviruses. *Curr. Top. Microbiol. Immunol.*, **114**, 153–75
13. Lehmann-Grube, F. (1977). Lymphocytic choriomeningitis virus carrier mice. Factors determining virus persistence. *Medicina* (Buenos Aires), **37** (Suppl. 3), 78–89

14. Moskophidis, D., Cobbold, S. P., Waldmann, H. and Lehmann-Grube, F. (1987). Mechanism of recovery from acute virus infection: treatment of lymphocytic choriomeningitis virus-infected mice with monoclonal antibodies reveals that Lyt-2$^+$ T lymphocytes mediate clearance of virus and regulate the antiviral antibody response. *J. Virol.*, **61**, 1867–74

15. Oldstone, M. B. A. and Dixon, F. J. (1967). Lymphocytic choriomeningitis: production of antibody by "tolerant" infected mice. *Science*, **158**, 1193–5

16. Benson, L. and Hotchin, J. (1969). Antibody formation in persistent tolerant infection with lymphocytic choriomeningitis virus. *Nature*, **222**, 1045–7

17. Oldstone, M. B. A., Tishon, A. and Buchmeier, M. J. (1983). Virus-induced immune complex disease: genetic control of Clq binding complexes in the circulation of mice persistently infected with lymphocytic choriomeningitis virus. *J. Immunol.*, **130**, 912–8

18. Crowle, A. J. and Hu, C. C. (1966). Split tolerance affecting delayed hypersensitivity and induced in mice by pre-immunization with protein antigens in solution. *Clin. Exp. Immunol.*, **1**, 323–35

19. Asherson, G. L. (1967). Antigen-mediated depression of delayed hypersensitivity. *Br. Med. Bull.*, **23**, 24–9

20. Owen, R. D. (1945). Immunogenetic consequences of vascular anastomoses between bovine twins. *Science*, **102**, 400–1

21. Burnet, F. M. and Fenner, F. (1949). *The Production of Antibodies*. 2nd Edn. (Melbourne: Macmillan Co.)

22. Billingham, R. E., Brent, L. and Medawar, P. B. (1953). 'Actively acquired tolerance' of foreign cells. *Nature*, **172**, 603–6

23. Cihak, J. and Lehmann-Grube, F. (1974). Persistent infection of mice with the virus of lymphocytic choriomeningitis: virus-specific immunological tolerance. *Infect. Immun.*, **10**, 1072–6

24. Popescu, M., Löhler, J. and Lehmann-Grube, F. (1977). Infectious lymphocytes in mice persistently infected with lymphocytic choriomeningitis virus. *Z. Naturforsch.*, **32c**, 1026–8

25. Popescu, M., Löhler, J. and Lehmann-Grube, F. (1979). Infectious lymphocytes in lymphocytic choriomeningitis virus carrier mice. *J. Gen. Virol.*, **42**, 481–92

26. Doyle, M. V. and Oldstone, M. B. A. (1978). Interactions between viruses and lymphocytes. I. *In vivo* replication of lymphocytic choriomeningitis virus in mononuclear cells during both chronic and acute viral infections. *J. Immunol.*, **121**, 1262–9

27. Traub, E. (1939). Epidemiology of lymphocytic choriomeningitis in a mouse stock observed for four years. *J. Exp. Med.*, **69**, 801–17

28. Tijerina, R., Löhler, J., Chaturvedi, U. C. and Lehmann-Grube, F. (1980). Infection of murine T lymphocytes with lymphocytic choriomeningitis virus: effect of age of mice on susceptibility. *Z. Naturforsch.*, **35c**, 1062–5

29. Lehmann-Grube, F., Tijerina, R., Zeller, W., Chaturvedi, U. C. and Löhler, J. (1983). Age-dependent susceptibility of murine T lymphocytes to lymphocytic choriomeningitis virus. *J. Gen. Virol.*, **64**, 1157–66

30. Raff, M. C. and Wortis, H. H. (1970). Thymus dependence of θ-bearing cells in the peripheral lymphoid tissues of mice. *Immunology*, **18**, 931–42

31. Popescu, M. and Ostrow, D. H. (1982). Multiplication of lymphocytic choriomeningitis virus in thymocytes during its persistence in mice. *J. Gen. Virol.*, **61**, 293–8

32. Bruns, M., Cihak, J., Müller, G. and Lehmann-Grube, F. (1983). Lymphocytic choriomeningitis virus. VI. Isolation of a glycoprotein mediating neutralization. *Virology*, **130**, 247–51

33. Löhler, J. and Lehmann-Grube, F. (1981). Immunopathologic alterations of lymphatic tissues of mice infected with lymphocytic choriomeningitis virus. I. Histopathologic findings. *Lab. Invest.*, **44**, 193–204

34. Cole, G. A., Gilden, D. H., Monjan, A. A. and Nathanson, N. (1971). Lymphocytic choriomeningitis virus: pathogenesis of acute central nervous system disease. *Fed. Proc.*, **30**, 1831–41

35. Allan, J. E. and Doherty, P. C. (1985). Consequences of cyclophosphamide treatment in murine lymphocytic choriomeningitis: evidence for cytotoxic T cell replication *in vivo*. *Scand. J. Immunol.*, **22**, 367–74

36. Accinni, L., Archetti, I., Branca, M., Hsu, K. C. and Andres, G. (1978). Tubulo-interstitial (TI) renal disease associated with chronic lymphocytic choriomeningitis viral infection in mice. *Clin. Immunol. Immunopathol.*, **11**, 395–405

37. Pollard, M. and Sharon, N. (1969). Immunoproliferative effects of lymphocytic chorio-meningitis virus in germfree mice. *Proc. Soc. Exp. Biol. Med.*, **132**, 242–6
38. Traub, E. (1962). Can LCM virus cause lymphomatosis in mice? *Arch. Ges. Virusforsch.*, **11**, 667–82
39. Skinner, H. H., Knight, E. H. and Lancaster, M. C. (1980). Lymphomas associated with a tolerant lymphocytic choriomeningitis virus infection in mice. *Lab. Anim.*, **14**, 117–21
40. Moskophidis, D. and Lehmann-Grube, F. (1984). The immune response of the mouse to lymphocytic choriomeningitis virus. IV. Enumeration of antibody-producing cells in spleens during acute and persistent infection. *J. Immunol.*, **133**, 3366–70
41. Oldstone, M. B. A. and Dixon, F. J. (1970). Persistent lymphocytic choriomeningitis viral infection. III. Virus-anti-viral antibody complexes and associated chronic disease following transplacental infection. *J. Immunol.*, **105**, 829–37
42. Moskophidis, D., Löhler, J. and Lehmann-Grube, F. (1986). Antiviral antibody production in parenchymatous organs of lymphocytic choriomeningitis virus carrier mice. Enumeration of antibody-producing cells and immunohistochemical investigation of mononuclear cell infiltrates. *Med. Microbiol. Immunol.*, **175**, 113–16
43. Moskophidis, D., Löhler, J. and Lehmann-Grube, F. (1986). Antiviral antibody-producing cells in parenchymatous organs during a persistent virus infection. *J. Exp. Med,* **165**, 705–19
44. Kabat, E. A., Freedman, D. A., Murray, J. P. and Knaub, V. (1950). A study of the crystalline albumin, γ-globulin and total protein in the cerebrospinal fluid of one hundred cases of multiple sclerosis and in other diseases. *Am. J. Med. Sci.*, **219**, 55–64
45. Heremans, J. F. (1968). Immunoglobulin formation and function in different tissues. *Curr. Top. Microbiol. Immunol.*, **45**, 131–203
46. Frydén, A., Link, H. and Norrby, E. (1978). Cerebrospinal fluid and serum immuno-globulins and antibody titers in mumps meningitis and aseptic meningitis of other etiology. *Infect. Immun.*, **21**, 852–61
47. Kinnman, J., Link, H. and Frydén, A. (1981). Characterization of antibody activity in oligoclonal immunoglobulin G synthesized within the central nervous system in a patient with tuberculous meningitis. *J. Clin. Microbiol.*, **13**, 30–5
48. Vandvik, B., Nilsen, R. E., Vartdal, F. and Norrby, E. (1982). Mumps meningitis: specific and non-specific antibody responses in the central nervous system. *Acta Neurol. Scand.*, **65**, 468–87
49. Forsberg, P. and Kam-Hansen, S. (1983). Immunoglobulin-producing cells in blood and cerebrospinal fluid during the course of aseptic meningoencephalitis. *Scand. J. Immunol.*, **17**, 531–7
50. Burke, D. S., Nisalak, A., Lorsomrudee, W., Ussery, M. A. and Laorpongse, T. (1985). Virus-specific antibody-producing cells in blood and cerebrospinal fluid in acute Japanese encephalitis. *J. Med. Virol.*, **17**, 283–92
51. Norrby, E., Link, H. and Olsson, J.-E. (1974). Measles virus antibodies in multiple sclerosis. Comparison of antibody titers in cerebrospinal fluid and serum. *Arch. Neurol.*, **30**, 287–92
52. Forghani, B., Cremer, N. E., Johnson, K. P., Fein, G. and Likosky, W. H. (1980). Comprehensive viral immunology of multiple sclerosis. III. Analysis of CSF antibodies by radioimmunoassay. *Arch. Neurol.*, **37**, 616–19
53. Vartdal, F., Vandvik, B., Michaelsen, T. E., Loe, K. and Norrby, E. (1982). Neurosyphilis: intrathecal synthesis of oligoclonal antibodies to *Treponema pallidum*. *Ann. Neurol.*, **11**, 35–40
54. Pedersen, N. Strandberg, Kam-Hansen, S., Link, H. and Mavra, M. (1982). Specificity of immunoglobulins synthesized within the central nervous system in neurosyphilis. *Acta Pathol. Microbiol. Immunol. Scand. Sect. C*, **90**, 97–104
55. Connolly, J. H. (1968). Additional data on measles virus antibody and antigen in subacute sclerosing panencephalitis. *Neurology*, **18**, 87–9
56. Tourtellotte, W. W., Ma, B. I., Brandes, D. B., Walsh, M. J. and Potvin, A. R. (1981). Quantification of *de novo* central nervous system IgG measles antibody synthesis in SSPE. *Ann. Neurol.*, **9**, 551–6
57. Weil, M. L., Itabashi, H. H., Cremer, N. E., Oshiro, L. S., Lennette, E. H. and Carnay, L. (1975). Chronic progressive panencephalitis due to rubella virus simulating subacute sclerosing panencephalitis. *N. Engl. J. Med.*, **292**, 994–8

58. Vandvik, B., Weil, M. L., Grandien, M. and Norrby, E. (1978). Progressive rubella virus panencephalitis: synthesis of oligoclonal virus-specific IgG antibodies and homogeneous free light chains in the central nervous system. *Acta Neurol. Scand.*, **57**, 53–64
59. Griffin, D. E., Narayan, O., Bukowski, J. F., Adams, R. J. and Cohen, S. R. (1978). The cerebrospinal fluid in visna, a slow viral disease of sheep. *Ann. Neurol.*, **4**, 212–8
60. Nathanson, N., Petursson, G., Georgsson, G., Palsson, P. A., Martin, J. R. and Miller, A. (1979). Pathogenesis of visna. IV. Spinal fluid studies. *J. Neuropathol. Exp. Neurol.*, **38**, 197–208
61. Prineas, J. W. (1979). Multiple sclerosis: presence of lymphatic capillaries and lymphoid tissue in the brain and spinal cord. *Science*, **203**, 1123–5
62. Gerhard, W., Taylor, A., Sandberg-Wollheim, M. and Koprowski, H. (1985). Longitudinal analysis of three intrathecally produced immunoglobulin subpopulations in an MS patient. *J. Immunol.*, **134**, 1555–60
63. Walsh, M. J. and Tourtellotte, W. W. (1986). Temporal invariance and clonal uniformity of brain and cerebrospinal IgG, IgA, and IgM in multiple sclerosis. *J. Exp. Med.*, **163**, 41–53

7
Differentiation-linked Susceptibility of Nervous and Muscular Tissue of Mice to Infection with Moloney Murine Leukaemia Virus

J. LÖHLER

INTRODUCTION

Like many other infectious micro-organisms viruses frequently replicate preferentially in selected differentiated cell populations. Factors that determine this tissue tropism are manifold and have to be ascertained for each virus–cell system[1,2]. For some time it was thought that expression of murine leukaemia viruses (MuLV) in particular was confined to lymphatic tissue[3-7]. However, Gardner and coworkers[8] isolated from wild mice ecotropic murine leukaemia viruses that, after prolonged latent periods, induced progressive hind limb paralysis in wild *Mus musculus* or susceptible laboratory mice. Subsequent investigations[9-11] revealed that besides leukaemogenic potential these viruses cause a spongiform encephalomyelopathy that is not accompanied by immunopathology or inflammation. Recently, further murine leukaemia viruses inducing CNS disease have been described. They were derived as temperature-sensitive mutants of Moloney murine leukaemia virus (Mo-MuLV)[12-14], isolates of rat-passaged Friend murine leukaemia virus[15], and *in vitro* constructed recombinant viruses[16]. Incidence and latency of paralysis are age-dependent[8,13,14,17,18], and resistance to paralysis is complete by 10 days of age. This observation suggests that a close relationship exists between the stage of CNS development and susceptibility to ecotropic MuLV of wild-mouse origin or to ts mutants, respectively. To find out whether such a developmentally controlled susceptibility is only valid for these particular virus–host relationships or is valid as a general rule for the interaction of MuLV with the mouse CNS, a study was undertaken to analyse virus expression in nervous and also muscular tissue of adult mice that had been exposed at different pre- and postnatal developmental stages to wild type Mo-MuLV which is not paralytogenic. For this investigation inbred mouse strains carrying a single

85

Mo-MuLV proviral copy stably integrated at different chromosomal positions into the mouse genome seemed to be optimally suited. These Mov-strains have been generated by Jaenisch and his colleagues either by infection of mouse embryos with Mo-MuLV or injection of embryos with cloned proviral DNA[19-21]. In some of the Mov-strains infectious virus is activated at different stages of development while in other Mov-strains the proviral genome remains inactive life-long. Since neuronal and muscular tissues of adult mice are non-permissive for MuLV the Mov-strains provide a very good model system to study the effect of differentiation of these tissues on virus replication and expression.

EXPRESSION OF MOLONEY MURINE LEUKAEMIA VIRUS RNA AND VIRUS SPECIFIC POLYPEPTIDES IN CNS AND MUSCLE TISSUE IN MICE OF MOV-STRAINS AND EXOGENOUSLY INFECTED MICE

To test the susceptibility of neuronal and muscle tissue for infection with Mo-MuLV, four Mov strains were selected in which the integrated provirus is activated at different stages of development (Table 7.1)[23]. In Mov-13 and Mov-9 mice, virus is expressed during the last third of the embryonal period, namely on days 16 and 18, respectively. In Mov-1 mice, viraemia can be detected on day 7 after birth whereas in Mov-6 mice the provirus remains inactive life-long. In addition to the Mov-strains, C57BL/6J mice infected intravenously with Mo-MuLV at the age of 8 weeks were employed and, to follow virus spread to CNS and musculature during development from mid-gestation up to adulthood, embryos were infected by micro-injection of Mo-MuLV directly through the uterus wall at day 8.5 of gestation.[24] For the

Table 7.1   Expression of Moloney murine leukaemia virus RNA and polypeptides in CNS and muscle tissue of adult mice of strains with germ line-integrated retrovirus and C57BL/6J (B6) infected as day 8 embryos or adult animals

Mo-MuLV genetic locus	Time of virus activation	Viral RNA			Viral proteins (gp70, p30)		
		CNS	Muscle	Other organs	CNS	Muscle	Other organs
Mov-1	7 days after birth	+	−	+	+	−	+
Mov-6	No activation	−	−	−	−	−	−
Mov-9	Day 18 of gestation	+	+	+	+	+	+
Mov-13	Day 16 of gestation	+	+	+	+	+	+
	Exogenous, intrauterine infection of day-8 B6 embryos	+	+	+	+	+	+
	Intravenous infection of 8-week old B6 mice	−	−	+	−	−	+

Mov substrains, carrying a single Moloney leukaemia proviral copy as Mendelian determinant, were derived by exposing mouse embryos to infectious virus at different developmental stages. Mov-1, Mov-6, Mov-9 were derived from virus-infected pre-implantation embryos[19,21] and Mov-13 from an embryo micro-injected with Mo-MuLV at mid-gestation[24]

86

demonstration of Mo-MuLV transcription and translation in the CNS and musculature, *in situ* hybridization and the peroxidase–antiperoxidase (PAP) technique were employed, respectively; both methods have been described previously[25]. The nucleic acid probe consisted of a ^{3}H-labelled Mo-MuLV cDNA and the polyclonal anti-Mo-MuLV antisera were specific for the viral proteins gp70 and p30. The anti-p30 antiserum was monospecific for the Mo-MuLV virus gag protein and did not cross-react with endogenous retrovirus proteins.

Table 7.1 summarizes the findings with mice of the various Mov-strains and B6 mice infected exogenously as embryos or adults. Different patterns of virus expression can be distinguished according to the age of the animal at which virus becomes activated or is injected. Whereas infection of midgestation embryos leads to an overall expression of Mo-MuLV in the whole organism, infection of adult mice results in a predominantly lymphotropic virus expression.

Correspondingly, occurrence of infectious virus in Mov-1, Mov-9 and Mov-13 mice at ages between these extremes is associated with a specific pattern of virus expression. A more detailed description for CNS and musculature will be presented below with particular reference to mice of Mov-1 and Mov-13 substrains, and B6 mice infected *in utero* at midgestation.

In the forebrain of adult Mov-1 mice, which activate provirus 7 days after birth, few astrocytes of white matter structures such as the corpus callosum (Figure 7.1a) and the anterior and posterior commissures expressed viral components. More numerous positive astrocytes and also neurons were detected in the medulla oblongata (Figure 7.1b). In Mov-1 mice, no virus expression was observed in muscle tissue. Adult Mov-13 mice, which activate Mo-MuLV provirus on day 16 of gestation, exhibited an entirely different pattern of expression. In contrast to Mov-1 mice, medulla oblongata and spinal cord were almost negative for viral RNA and polypeptides. However, a distinct virus expression became visible in the cerebellum (Figure 7.2a,b), hippocampus, bulbus olfactorius and corpus callosum. In these brain regions both neuronal and glial cells were involved in virus expression. In addition to astrocytes, positive oligodendrocytes could be identified in the corpus callosum. Virus-specific transcription and translation was also detectable in cells of skeletal, cardiac and smooth muscle tissue of Mov-13 mice (Figure 7.2c,d). Not only were viral RNA and structural proteins present in Mov-13 CNS and muscle tissue, preliminary results from electron microscopical investigations also disclosed mature type C virus particles. In contrast to Mov-1 and Mov-13 mice, it was not possible to recognize virus expression in the CNS and musculature of B6 mice infected at the age of 8 weeks; in these animals virus expression was restricted to lymphatic tissues.

The differential involvement of CNS structures and muscle tissue of Mov-1 and Mov-13 mice in virus expression suggested that permissiveness for productive infection with Mo-MuLV depends on the state of cell differentiation. Further evidence was obtained by studying the spread and distribution of virus in B6 mice that had been directly infected as mid-gestation embryos. Four days after injection of virus into day 8.5 embryos, multiple foci of cells expressing viral RNA were detected in CNS cells as well as in cells of other

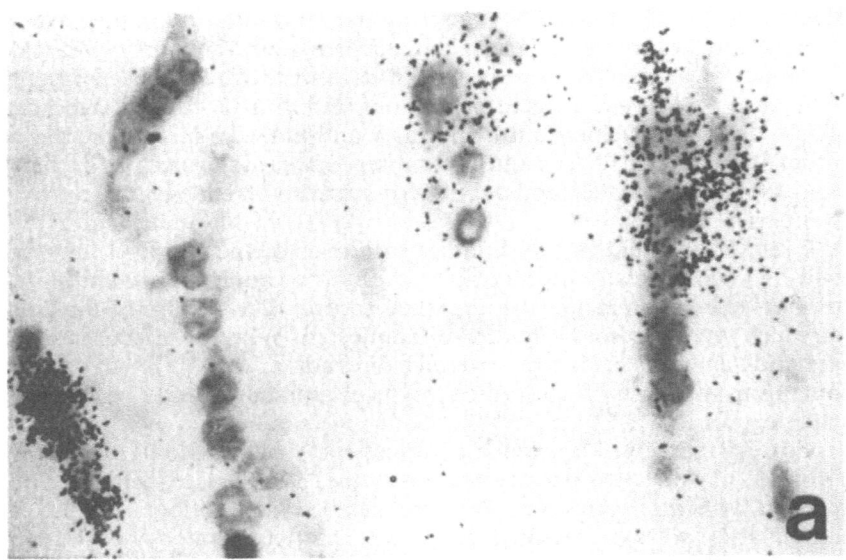

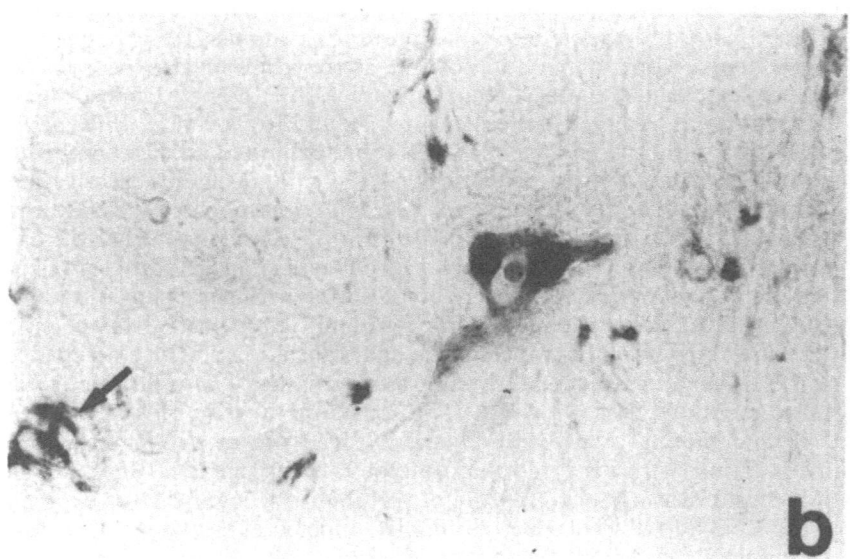

Figure 7.1 The Mov-1 strain activates the Mo-MuLV 7 days after birth. (a) Expression of viral RNA in a few glial cells of the fimbria hippocampi (*in situ* hybridization). Immunolabelling for viral polypeptides enables identification of these positive cells as astrocytes. (b) In Mov-1 mice expression of viral p30 in neurons (arrow) is confined to a few cells of the medulla oblongata and spinal cord. The arrow head points to a positive astrocyte. (PAP technique)

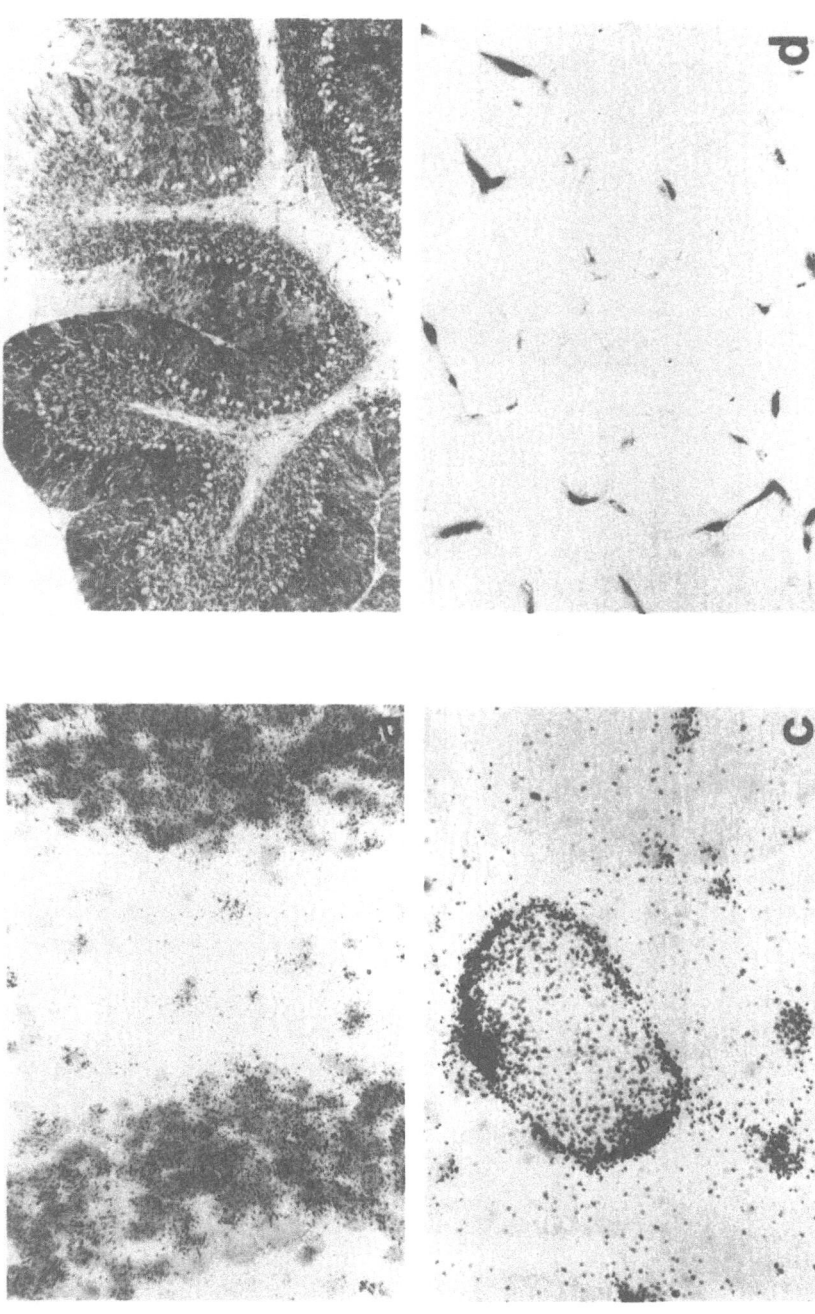

Figure 7.2 The Mov-13 strain activates the Mo-MuLV on day 16 of gestation. (a) Expression of viral RNA in most granule and glial cells of the cerebellum (*in situ* hybridization). (b) Immunostaining for p30 shows widespread protein expression throughout the cortex and white matter of the cerebellum, however, remarkably the Purkinje cells are spared. (PAP technique.) (c) *In situ* hybridization of viral RNA in the perinuclear sarcoplasm of skeletal muscle fibres, (d) A comparable pattern is obtained by immunolabelling of viral gp70. (PAP technique)

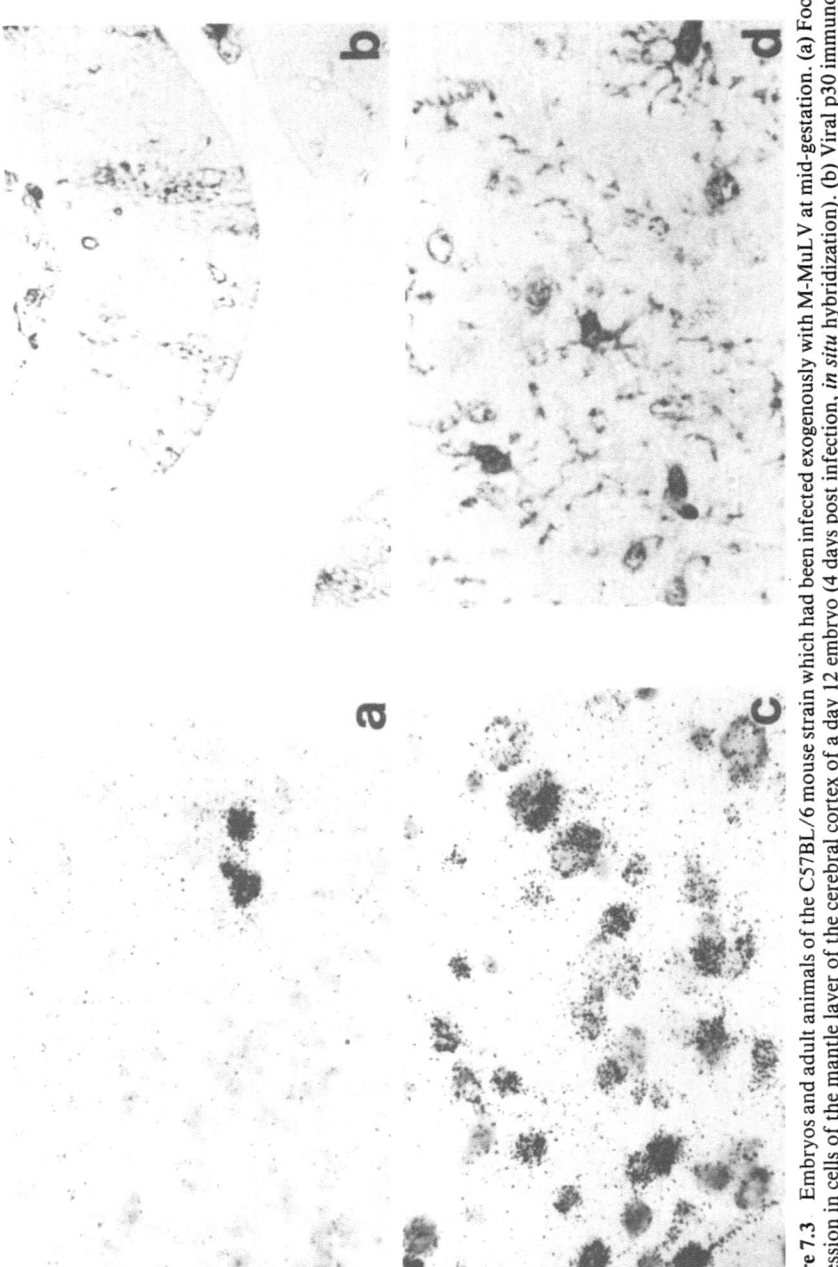

Figure 7.3 Embryos and adult animals of the C57BL/6 mouse strain which had been infected exogenously with M-MuLV at mid-gestation. (a) Focal RNA expression in cells of the mantle layer of the cerebral cortex of a day 12 embryo (4 days post infection, *in situ* hybridization). (b) Viral p30 immunostained in numerous cells of the wall of a telencephalic vesicle of a day 13 embryo. (PAP technique) (c) Viral RNA expression in neuronal and glial cells of the cerebral cortex of C57BL/6 mouse infected as day 8.5 embryo. (d) Astrocytes and oligodendrocytes of the corpus striatum are immunolabelled for p30. (PAP technique)

tissues. For example, in the telencephalic anlage of day 12 embryos RNA was expressed in cells of the cortical mantle layer (Figure 7.3a), and immuno-staining of viral polypeptides revealed foci of labelled cells (Figure 7.3b). Futhermore, at the same time spinal ganglia, too, could be detected expressing virus. During the following days the infection spread throughout the developing CNS. In adult B6 mice derived from infected embryos, a similar pattern of virus expression was found as in those Mov-13 mice that activate virus on day 16 of gestation, but the intensity of RNA and protein expression was much more pronounced in the former, and additional brain structures were involved. Most importantly, in B6 mice virus expression was also seen in glial and neuronal cells of the cerebral cortex, striatum, thalamus and hypothalamus (Figure 7.3c,d) which, in Mov-13 mice, were free of viral antigen and nucleic acid. A further striking observation in the B6 mice was that expression of viral proteins also occurred in myelin sheaths of, for example, myelin fibre bundles of the corpus striatum.

PUTATIVE ROLE OF CELL PROLIFERATION FOR PERMISSIVE MURINE RETROVIRUS INFECTION OF CNS AND MUSCLE CELLS

The results from studying Mo-MuLV expression in the brains of Mov-1, Mov-13, and prenatally infected B6 mice are summarized in Table 7.2. They strongly suggest that a relationship exists between susceptibility to Mo-MuLV and the stage of differentiation of the cellular elements of the developing CNS. In particular, the selective expression of virus in cere-bellum, hippocampus and bulbus olfactorius of Mov-13 mice directs our attention to the fact that in these regions proliferation of neuronal and glial cells starts on day 14 of gestation[26-28], which coincides with the time of

Table 7.2 The differential expression of Moloney murine leukaemia virus RNA and proteins in CNS regions of adult Mov-1 and Mov-13 mice, and in adult C57BL/6 mice infected on day 8.5 of embryogenesis

Mouse strain	Time of virus activation or infection	CNS region
Mov-1	7 days after birth	Astrocytes and neurons of medulla oblongata Few astrocytes of corpus callosum, and anterior and posterior commissures
Mov-13	Day 16 of gestation	Glial and neuronal cells of cerebellum, hippocampus, and bulbus olfactorius Astrocytes of corpus callosum and anterior and posterior commissures
C57BL/6/J	Intrauterine on day 8.5 of gestation	Astrocytes, oligodendrocytes, and neurons of all CNS regions Spinal ganglia

virus activation, namely day 16. On the other hand earlier exposure of Mo-MuLV to immature CNS cells during the decisive period day 10 to day 13, when major brain structures such as the hemispheres, striatum and thalamus are forming, leads to persistent virus expression in these regions in adult mice. Virus activation at day 7 after delivery is associated with a pattern of virus expression that is confined to astroglial cells of distinct white matter structures, to neuronal and glial cells of the medulla oblongata, and eventually the spinal cord. At this developmental stage morphogenesis and differentiation of the CNS are complete with the exception of the so-called myelination gliosis[42]. These observations prompt the question of what the reasons are for this differentiation-linked pattern of virus expression. Is it the programmed state of cell proliferation that renders specific CNS regions permissive for virus? It is known that Mo-MuLV does. not replicate in cells that are in the G0/G1 phase of the cell cycle[29–31]. But how then do neuronal cells of adult mice which are arrested in the G0 phase retain their capability for virus replication?

One could speculate that an active cell cycle phase is necessary for the virus to infect the cell and that cellular proliferation favours virus replication but is not essential for virus production when (by some unknown mechanism) chronically infected cells uncouple the block on MuLV that is usually imposed by cell cycle arrest. Recently, Gloger and coworkers[32] have shown that treatment of chronically Mo-MuLV infected cultivated cells with butyric acid (which arrests cells at the G0/G1 phase) did not inhibit virus production. Furthermore, the inhibition caused by other methods was overcome by butyric acid. Perhaps dissociation of virus replication from cell proliferation may also occur in CNS cells arrested in the G0 phase, provided infection is initiated when cells are still proliferating and differentiating. Some evidence for such a mechanism functioning in the age-related restriction of hind-limb paralysis caused by wild-mouse ectropic leukaemia virus has been provided by the work of Brooks and co-authors[33,34]. In astrocyte cultures established from embryos or newborn mice of different ages (embryo day 17, newborn day 1 and 10) synthesis of proteins of neurotropic retrovirus decreased with increasing stages of cellular development although the virus adsorbed more effectively to astrocytes obtained from 10-day-old newborn mice than from day 17 embryos. Immunological reactions were proposed by Hoffman and coworkers[18] to play a role in age-related resistance to paralysis after infection with wild-mouse (Cas-Br-M) MuLV.

Of course, other cellular and viral factors exist which are important in regulating the susceptibility of distinct cell types for retrovirus replication and ensuing pathological effects. The presence of an appropriate receptor for the envelope glycoprotein (gp70) is essential for the penetration of murine leukaemia viruses into cells[35,36], thus restricting permissive infection at the cell surface. Equally important are the genetic loci of the host, e.g. the Fv-1 locus of the mouse[37], which constitute an intracellular restriction by impairing viral integration after penetration. Furthermore, evidence has been put forward for the existence of certain control sequences in the long terminal repeats which confer tissue and disease specificity[16,38–41]. Using chimeric viruses that harbour the *gag-pol-env* sequences of the neurotropic wild-mouse retrovirus Cas-BR-E and either the non-neurotropic amphotropic 4070-A-murine leukaemia virus or the lymphotropic Mo-MuLV, DesGroseillers and co-authors[16] found the primary determinant of paralysis to be mapped within a 3.9 kb pair *pol-env* fragment. Long terminal repeat sequences seem to influence the incidence of the disease and the distribution pattern of the neuropathological lesions.

Considering the various factors that influence and regulate incidence, tissue tropism and selective disease induction of MuLV infections, one has to conclude that at the moment a definite explanation of the differentiation-linked susceptibility of nervous and muscular tissue to MuLV infection cannot be presented. However, by extending investigations to prenatal stages of development, the findings presented in this report show for the first time that the developmentally restricted susceptibility of CNS and musculature to murine retrovirus infection is not an exclusive feature of paralytogenic wild-mouse leukaemia virus and neurotropic ts mutants of Mo-MuLV but is generally valid for the interaction of MuLV with CNS and musculature. They demonstrate, furthermore, that the wild-type Mo-MuLV can also become neurotropic without causing disease and that the stage of differentiation is of decisive importance for the induction of permissive retrovirus infections of the CNS. These observations may also provide new insight into the preferential infection of the CNS of newborn children infected by the human immune deficiency virus (HIV) *in utero* or shortly after birth.

Acknowledgements

I would like to thank Dr Rudolf Jaenisch for generously providing the Mov-strains. The micro-injection of mid-gestation embryos by Dr Heidi Stuhlmann is gratefully acknowledged. Part of the *in situ* hybridization procedure was performed by Dr Iva Simon. I thank Helga Siegel for the preparation of the manuscript.

The Heinrich-Pette-Institut is financially supported by Freie und Hansestadt Hamburg and Bundesministerium für Jugend, Familie, Frauen und Gesundheit, Bonn.

References

1. Maltzman, W. and Levine, A. J. (1981). Viruses as probes for development and differentiation. *Adv. Virus Res.*, 26, 65–116
2. Levine, A. J. (1984). Viruses and differentiation: the molecular basis of viral tissue tropisms. In Notkins, A. L. and Oldstone, M. B. A. (eds.) *Concepts in Viral Pathogenesis.* pp. 130–4. (New York: Springer-Verlag)
3. Rowe, W. P. and Pincus, T. (1972). Quantitative studies of naturally occurring murine leukemia virus infection of AKR mice. *J. Exp. Med.*, 135, 429–36
4. Hilgers, J., Declève, A., Galesloot, J. and Kaplan, H. S. (1974). Murine leukemia virus group-specific antigen expression in AKR mice. *Cancer Res.*, 34, 2553–61
5. Declève, A., Travis, M., Weissman, J., Lieberman, M. and Kaplan, H. (1975). Focal infection and transformation *in situ* of thymus cell subclasses by a thymotropic murine leukemia virus. *Cancer Res.*, 35, 3585–95
6. Gisselbrecht, S., Blaineau, Ch., Hurot, M. A., Pozo, F. and Levy, P. (1978). Prevalence of non-T cells in the replication of the N-tropic, type C virus of young AKR mice. *Cancer Res.*, 38, 939–41

7. Jaenisch, R. (1979). Moloney leukemia virus gene expression and gene amplification in preleukemic and leukemic BALB/Mo mice. *Virology*, **93**, 80–90
8. Gardner, M. B., Henderson, B. E., Officer, J. E., Rongey, R. W., Parker, J. C., Oliver, C., Estes, J. D. and Huebner, R. J. (1973). A spontaneous lower motor neuron disease apparently caused by indigenous type C RNA virus in wild mice. *J. Natl. Cancer Inst.*, **51**, 1243–54
9. Andrews, J. M. and Gardner, M. B. (1974). Lower motor neuron degeneration associated with type C RNA virus infection in mice: neuropathological features. *J. Neuropathol. Exp. Neurol.*, **33**, 285–307
10. Oldstone, M. B. A., Lampert, P. W., Lee, S. and Dixon, F. J. (1977). Pathogenesis of the slow disease of the central nervous system associated with WM 1504 E virus. I. Relationship of strain susceptibility and replication to disease. *Am. J. Pathol.*, **88**, 193–212
11. Brooks, B. R., Swarz, J. R. and Johnson, R. T. (1980). Spongiform polioencephalomyelopathy caused by a murine retrovirus. *Lab. Invest.*, **43**, 480–6
12. McCarter, J. A., Ball, J. K. and Frei, J. V. (1977). Lower limb paralysis induced in mice by a temperature-sensitive mutant of Moloney leukemia virus. *J. Natl. Cancer Inst.*, **59**, 179–83
13. Wong, P. K. Y., Soong, M. M., MacLeod, R., Gallick, G. and Yuen, P. H. (1983). A group of temperature-sensitive mutants of Moloney leukemia virus which is defective in cleavage of *env* precursor polypeptide in infected cells also induces hind limb paralysis in newborn CFW/D mice. *Virology*, **125**, 513–8
14. Bilello, J. A., Pitts, O. M. and Hoffman, P. M. (1986). Characterization of a progressive neurodegenerative disease induced by a temperature-sensitive Moloney murine leukemia virus infection. *J. Virol.*, **59**, 234–41
15. Kai, K. and Furuta, T. (1984). Isolation of paralysis-inducing murine leukemia viruses from Friend virus passage in rats. *J. Virol.*, **50**, 970–3
16. DesGroseillers, L., Rassart, E., Robitaille, Y. and Jolicoeur, P. (1985). Retrovirus-induced spongiform encephalopathy: the 3′-end long terminal repeat-containing viral sequences influence the incidence of the disease and the specificity of the neurological syndrome. *Proc. Natl. Acad. Sci. USA*, **82**, 8818–22
17. Hoffman, P. M., Ruscetti, S. K. and Morse, H. C. (1981). Pathogenesis of paralysis and lymphoma associated with a wild mouse retrovirus infection, Part 1. Age- and dose-related effects in susceptible laboratory mice. *J. Neuroimmunol.*, **1**, 275–85
18. Hoffman, P. M., Robbins, D. S. and Morse, H. C. (1984). Role of immunity in age-related resistance to paralysis after murine leukemia virus infection. *J. Virol.*, **52**, 734–8
19. Jaenisch, R. (1976). Germ line integration and Mendelian transmission of the exogenous Moloney leukemia virus. *Proc. Natl. Acad. Sci. USA*, **73**, 1260–4
20. Jähner, D. and Jaenisch, R. (1980). Integration of Moloney leukemia virus into the germ line of mice: correlation between genotype and virus activation. *Nature*, **287**, 456–8
21. Jaenisch, R., Jähner, D., Nobis, P., Simon, I., Löhler, J., Harbers, K. and Grotkopp, D. (1981). Chromosomal position and activation of retroviral genomes inserted into the germ line of mice. *Cell*, **24**, 519–29
23. Jaenisch, R., Breindl, M., Harbers, K., Jähner, D. and Löhler, J. (1985). Retroviruses and insertional mutagenesis. *Cold Spring Harbor Symp. Quant. Biol.*, **50**, 439–45.
24. Jaenisch, R. (1980). Retroviruses and embryogenesis: micro-injection of Moloney leukemia virus into mid-gestation mouse embryos. *Cell*, **19**, 181–8
25. Simon, I., Löhler, J. and Jaenisch, R. (1982). Virus-specific transcription and translation in organs of BALB/Mo mice: comparative study using quantitative hybridization, *in situ* hybridization, and immunocytochemistry. *Virology*, **120**, 106–21
26. Altman, J. and Das, G. D. (1966). Autoradiographic and histological studies of postnatal neurogenesis. I. A longitudinal investigation of the kinetics, migration and transformation of cells incorporating tritiated thymidine in neonate rats, with special reference to postnatal neurogenesis in some brain regions. *J. Comp. Neurol.*, **126**, 337–90
27. Altman, J. (1966). Proliferation and migration of undifferentiated precursor cells in the rat during postnatal gliogenesis. *Exp. Neurol.*, **16**, 263–78
28. Schlessinger, A. R., Cowan, W. M. and Gottlieb, D. J. (1975). An autoradiographic study of the time of origin and the pattern of granule cell migration in the dentate gyrus of the rat. *J. Comp. Neurol.*, **159**, 149–76

29. Paskind, M. P., Weinberg, R. A. and Baltimore, D. (1975). Dependence of Moloney murine leukemia virus production on cell growth. *Virology*, **67**, 242–8
30. Sherton, C. C., Evans, L. H., Polonoff, E. and Kabat, D. (1976). Relationship of Friend murine leukemia virus production to growth and hemoglobin synthesis in cultured erythroleukemia cells. *J. Virol.*, **19**, 118–25
31. Balazs, J. and Caldarella, J. (1981). Retrovirus gene expression during the cell cycle. I. Virus production, synthesis, and expression of viral proteins in Rauscher murine leukemia virus-infected mouse cells. *J. Virol.*, **39**, 792–9
32. Gloger, J., Arad, G. and Panet, A. (1985). Regulation of Moloney murine leukemia virus replication in chronically infected cells arrested at the G0/G1 phase. *J. Virol.*, **54**, 844–50
33. Brooks, B. R., Gossage, J. and Johnson, R. T. (1981). Age-dependent *in vitro* restriction of mouse neurotropic retrovirus replication in central nervous system-derived cells from susceptible Fv-1^{nn} mice. *Trans. Am. Neurol. Assoc.*, **106**, 238–41
34. Brooks, B. R., Priester, E. and Pal, B. (1984). Age-related restriction of murine neurotropic retrovirus infection of astrocytes *in vitro* is due to decreased synthesis of specific virus proteins. *Neurology*, **34** (Suppl. 1), 178–9
35. DeLarco, J. and Todaro, G. J. (1976). Membrane receptors for murine leukemia viruses: characterization using purified viral envelope glycoprotein, gp71. *Cell*, **8**, 365–71
36. Johnson, P. A. and Rosner, M. R. (1986). Characterization of murine-specific leukemia virus receptor from L cells. *J. Virol.*, **58**, 900–8
37. Hartley, J. W., Rowe, W. P. and Huebner, R. J. (1970). Host-range restrictions of murine leukemia viruses in mouse embryo cell cultures. *J. Virol.*, **5**, 221–5
38. Rosen, C. A., Haseltine, W. A., Lenz, J., Ruprecht, R. and Cloyd, M. W. (1985). Tissue selectivity of murine leukemia virus infection is determined by long terminal repeat sequences. *J. Virol.*, **55**, 862–6
39. Fan, H., Mittal, S., Chute, H., Chao, E. and Pattengale, P. K. (1986). Rearrangements and insertions in the Moloney murine leukemia virus long terminal repeat alter biological properties *in vivo* and *in vitro*. *J. Virol.*, **60**, 204–14
40. Bösze, Z., Thiesen, H.-J. and Charnay, P. (1986). A transcriptional enhancer with specificity for erythroid cells is located in the long terminal repeat of the Friend murine leukemia virus. *EMBO J.*, **5**, 1615–23
41. Stocking, C., Kollek, R., Bergholz, U. and Ostertag, W. (1986). Point mutations in the U3 region of the long terminal repeat of Moloney murine leukemia virus determine disease specificity of the myeloproliferative sarcoma virus. *Virology*, **153**, 145–9
42. Altman, J. (1966). Autoradiographic and histological studies of postnatal neurogenesis. II. A longitudinal investigation of the kinetics, migration and transformation of cells incorporating tritiated thymidine in infant rats, with special reference to postnatal neurogenesis in some brain regions. *J. Comp. Neurol.*, **128**, 431–74

8
Antigenic Phenotype and Experimental Herpes Simplex Virus Infection of Cultured Human Fetal Neural Cells

P. G. E. KENNEDY

INTRODUCTION

Human neural cell cultures have proved to be a useful tool in studying the developmental biology of the human nervous system[1,2]. These tissue culture systems have also been used in a variety of ways to investigate the pathogenesis of various neurological diseases in which immunopathological mechanisms may be important[3-5] and the susceptibility of human neural cells to infection with viruses[6,7]. We have favoured the use of cultured human fetal neural cells in such studies because of their greater relevance to human disease compared with cell cultures derived from animal tissues, and also in view of the obvious ethical contraindications to studying such cells *in vivo*.

These studies have been possible because of the ease with which human fetal cells can be grown in culture, compared with adult cells, and the availability of cell-type-specific antigenic markers which can identify unambiguously all of the major neural cell types in such cultures. These cell markers were initially developed in dissociated cell cultures of neonatal rodent nervous tissues by Raff and his colleagues[8,9] and were subsequently shown to be effective in similar cultures prepared from aborted human fetal tissues[1,10]. Morphological analysis of such cellularly heterogeneous cultures has been shown to be an inadequate and often inaccurate method of cell identification; the ability of cell marker antibodies to identify cell types which may constitute only a small proportion of the total cell population has proved superior to morphological criteria alone[9,11].

There are two main ways in which human fetal neural cell cultures have been used in developmental studies of the human nervous system both of which have relevance to the subject of this symposium:

(1) Since these cells grow and change their antigenic phenotype in culture it has been possible to study their development *in vitro*, and the ontogeny of specific neural cell types can be determined;

(2) by virtue of their derivation from fetal tissues they can be used to study the susceptibility of developing human neural cells to infection with various viruses which cause human neurological diseases.

This chapter will describe studies which address both of these issues.

ANTIGENIC PHENOTYPE OF CULTURED HUMAN FETAL NEURAL CELLS

Initial studies in dissociated cell cultures

Our studies of cell marker expression in human neural cultures were initiated at University College, London, in 1979. It proved possible to apply almost all of the markers which had previously been used successfully in the rodent nervous system to human cells[1]. Two years later Dickson and colleagues[10] confirmed the efficacy of these various markers in human cultures and also used additional monoclonal antibody markers.

The experimental details of these studies have been described elsewhere[1, 2, 9] and will not be detailed here. In brief, dissociated cell cultures are produced by enzymatic digestion of nervous tissues obtained from 15–21 week old aborted human fetuses. The anatomical regions which have been used include optic nerve, cerebral cortex and hemispheres, spinal cord, dorsal root ganglia (DRG) and leptomeninges. After digestion the dissociated cultures are produced by trituration of tissues through hypodermic needles or Pasteur pipettes. The resulting cell suspensions are then seeded onto glass coverslips in culture medium for use in immunofluorescence experiments. These cultures are short-term and have generally been studied between 3 and 7 days in culture. The various cell-type-specific antibodies are applied to the cultures in indirect immunofluorescence assays. Cultures are stained live for surface staining and/or pre-fixed for visualization of intracellular antigens. Two-fluorochrome immunofluorescence has proved to be an invaluable tool in these studies. With this double-labelling technique cells are exposed to two different antibodies simultaneously or sequentially and their cell binding is then visualized with two different fluorochrome-conjugated antibodies[1, 9]. In this way the simultaneous expression of two antigens on the same cell can be demonstrated.

Astrocytes

The glial fibrillary acidic protein (GFAP)[12] is a protein associated with glial intermediate filaments and has been used extensively as a specific marker for astrocytes in cell cultures and tissue sections from a wide variety of animal species[9, 13, 14]. After acid–alcohol fixation of dissociated human optic nerve, spinal cord and brain cultures, cells with the morphological appearance of astrocytes were stained in a characteristic intracellular fibrillar pattern[1] (Figure 8.1). Some of these cells had the appearance of fibrous astrocytes with branching processes, others were large and flat, resembling protoplasmic astrocytes, while a few GFAP[+] (positive) cells had the morphological characteristics of both types. Although 30–70% of GFAP[+] cells also expressed the Thy-1-like glycoprotein[1, 15], in these initial studies they were not found to be labelled with any of the other cell markers (see below). Cultured adult brain cells have also been shown to express GFAP[16].

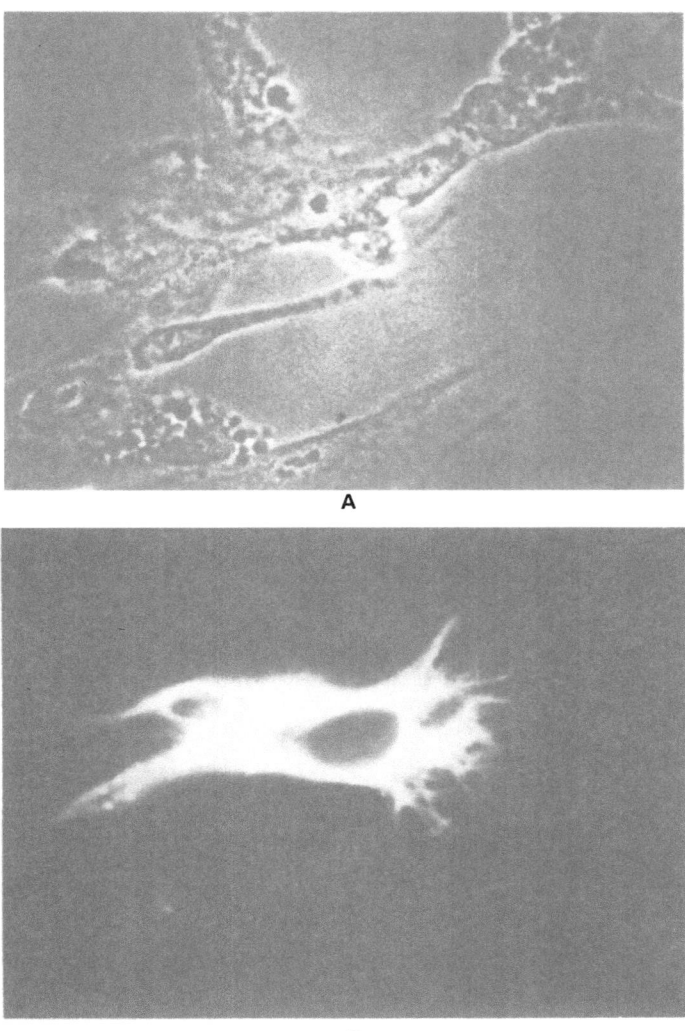

A

B

Figure 8.1 GFAP[+] astrocyte. Cells in culture of human optic nerve were fixed in acid–alcohol after 3 days and labelled with rabbit anti-GFAP serum followed by goat anti-rabbit Ig conjugated to fluorescein, and viewed with phase-contrast (A) or fluorescence (B) optics. Only one cell in the centre of the field is labelled. Note that the intracellular fibrillar structures detected by immunofluorescence are not visible on the phase-contrast image. (Figure reprinted with permission from *Laboratory Investigation* Vol. 43, pp. 342–51, 1980)

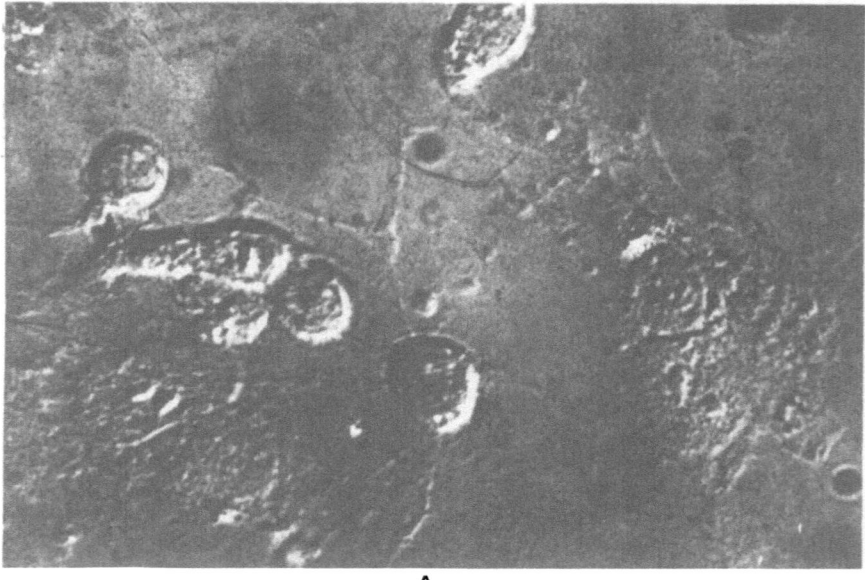

A

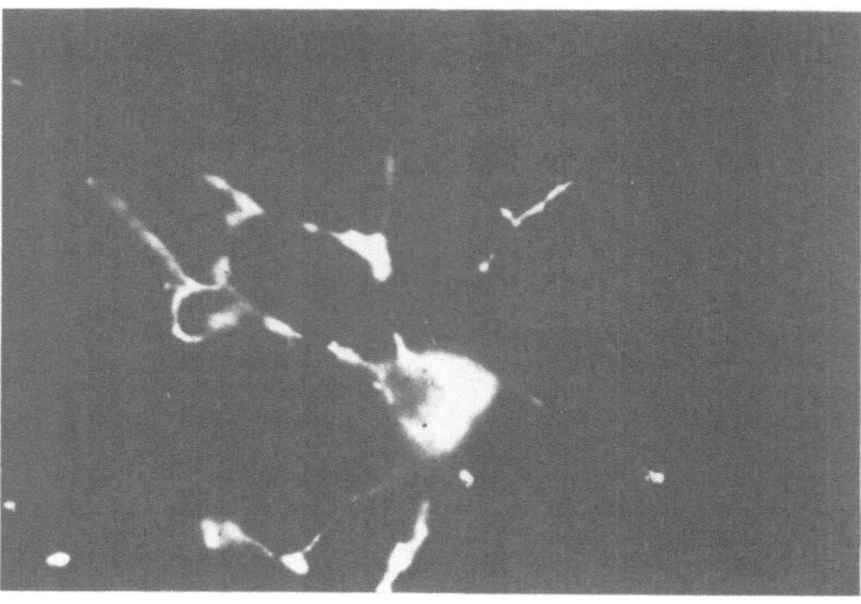

B

Figure 8.2 GC$^+$ oligodendrocyte. Cells in culture of human optic nerve were labelled after 3 days with rabbit anti-GC serum followed by goat anti-rabbit Ig conjugated to rhodamine, and viewed with Nomarski (A) or fluorescence (B) optics. Only one cell in the field shows surface labelling

Oligodendrocytes

The other major central nervous system (CNS) glial cell is the oligodendrocyte, the function of which is to form CNS myelin[17]. Galactocerebroside (GC), the major glycolipid in myelin[18], has been shown to be a specific cell surface antigenic marker for oligodendrocytes in CNS cultures obtained from a variety of species[8,11,19]. Human fetal oligodendrocytes could also be identified using antibodies to GC. The initial efficacy of this marker was proved in rat optic nerve cultures which do not contain neuronal cell bodies[8]. Similar experiments in human fetal optic nerve cultures showed that GC⁺ oligodendrocytes were not labelled with antibodies against GFAP; this observation had also been made in rat cultures. Some of the GC⁺ cells had multiple branching processes while others had rudimentary or no processes (Figure 8.2). Although approximately 50% of the GC⁺ cells in human spinal cord cultures also expressed intracellular myelin basic protein (MBP), GC⁺ cells in the optic nerve did not do so. Cultured adult human oligodendrocytes can also be identified with antibodies to GC[20].

Neurons

In early experiments using cell markers rat and human neurons in the CNS and peripheral nervous system (PNS) were identified by means of their surface labelling with tetanus toxin[21,22] (Figure 8.3). Although this is a useful marker it also binds non-specifically to cellular debris and it has been largely superseded by a variety of other markers including monoclonal antibodies recognizing specific neuronal cell populations[23,24]. Antibodies to specific neurofilament polypeptides have been particularly useful[25,26]. Two monoclonal antibodies which have been well studied in human cultures are A2B5 and A4. A2B5 appears to recognize a GQ-type ganglioside in chick neural retinal cultures[27], and was initially thought to be a specific neuronal marker in human central and peripheral neural cultures[10]. However, as will be described later it is now known that it is not neuron-specific. The A4 monoclonal antibody binds only to neurons in rat CNS cultures but not to PNS neurons[23]. In human cultures it has the same distribution in that it labels CNS neurons but not DRG neurons[10]. All of the tetanus toxin⁺ neurons in CNS and PNS cultures also expressed the Thy-1-like glycoprotein[1,10].

Schwann cells

Although cultured rat Schwann cells can be identified using antibodies to the rat neural antigen-1 (Ran-1)[28,29] unfortunately this marker is not effective in cultures derived from human or other non-rodent species. Partial success in identifying human Schwann cells was achieved by labelling DRG cultures with antibodies to the myelin-specific molecules GC and MBP (Figure 8.4). It was found that 25–30% of Schwann cell derived from 20-week-old fetuses expressed GC and MBP during the first 3 days in culture. However, these molecules were progressively lost with time in culture.

101

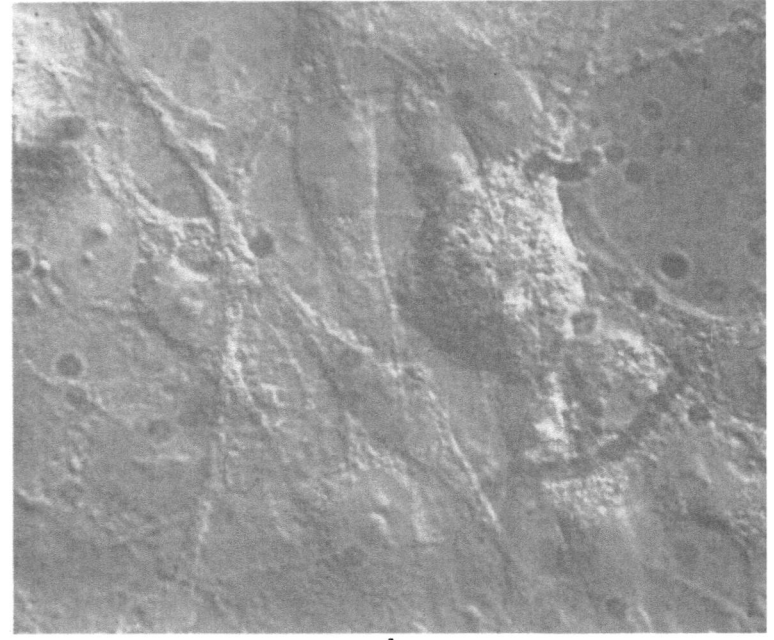

A

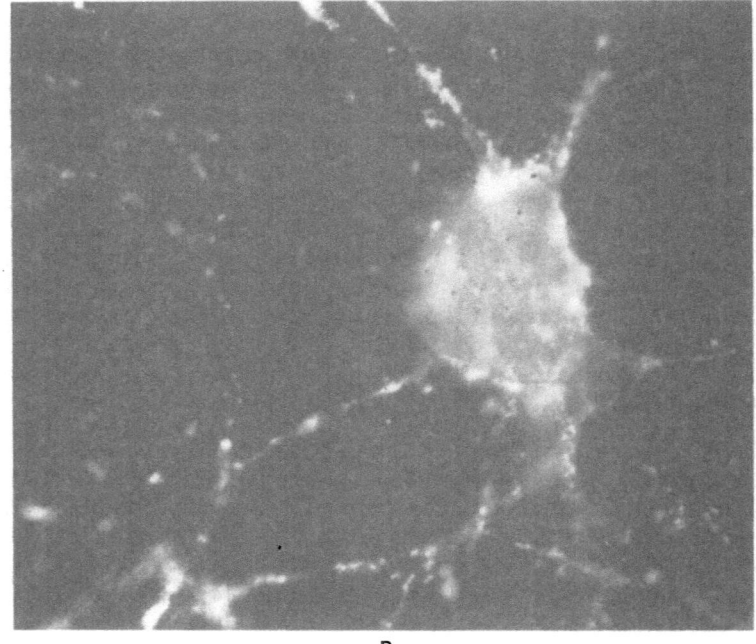

B

Fibroblasts and leptomeningeal cells

A large proportion of the cells derived from both rodent and human nervous tissues are large, flat GFAP[-] (negative) GC[-] fibroblastic cells which consist mainly of fibroblasts and leptomeningeal cells[1,9]. As in rodent cultures, it was found that these two cell populations could be distinguished from each other; although both cell types expressed the glycoprotein fibronectin (FN)[30], fibroblasts also expressed the Thy-1-like glycoprotein whereas leptomeningeal cells did not do so.

Macrophages

A small proportion (1–2%) of rat and human neural cell cultures contain macrophages. Although some of the cells are probably contaminating blood monocytes it is likely that a proportion of them are resident 'microglial cells'. As is the case in rodent cultures, these human macrophages can be identified by their phagocytic properties and their cell surface expression of Fc receptors. These receptors were not identified on other cell types in these cultures.

Recent developmental studies in human fetal optic nerve cultures

Several recent observations on glial cell development in the rat optic nerve have prompted us to extend our initial studies of antigen expression in human neural cell cultures. For example, Raff and his colleagues have shown that two distinct populations of astrocytes exist in the rodent CNS[31]. These have been termed Type 1 and Type 2 astrocytes and broadly correspond to protoplasmic and fibrous astrocytes respectively. Type 1 astrocytes have a GFAP[+] A2B5[-] phenotype whereas Type 2 astrocytes are GFAP[+] A2B5[+], and the properties of these different cell types have been shown to differ in a variety of ways[31]. Further, it has been demonstrated that Type 2 astrocytes and GC[+] oligodendrocytes in cultures of neonatal rat optic nerve develop from a common glial progenitor cell (0-2A cell)[32]. This bipotential progenitor cell is A2B5[+] and can be induced to develop into a Type 2 astrocyte if cultured in the presence of fetal calf serum (FCS) or into an oligodendrocyte if cultured in the absence of serum[32]. We therefore carried out a series of experiments to determine whether human optic nerve cells in culture share similar properties to the corresponding rodent cells.

Astrocyte heterogeneity in human fetal optic nerve cultures

In cultures of neonatal rat optic nerve it had been found that GFAP[+] astrocytes could be divided into two populations according to their labelling with the A2B5 monoclonal antibody. The Type 1 astrocytes (GFAP[+] A2B5[-] phenotype) were large flat cells whereas the Type 2 astrocytes (GFAP[+] A2B5[+]

Figure 8.3 Tetanus toxin[+] neuron. Cells in human DRG cultures were labelled with tetanus toxin, followed by mouse anti-tetanus toxin serum, followed by goat anti-mouse Ig conjugated to rhodamine and viewed with Nomarksi (A) or fluorescence (B) optics. Only the neuron and its processes are labelled

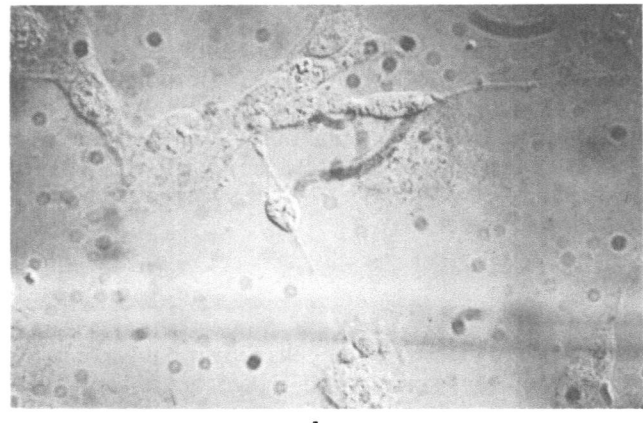

A

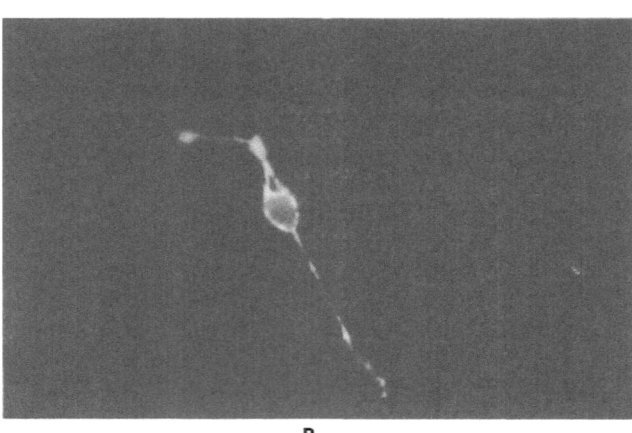

B

Figure 8.4 GC⁺ Schwann cell. After 2 days cells in human DRG cultures were incubated with rabbit anti-GC followed by goat anti-rabbit Ig conjugated to rhodamine and viewed with Nomarksi (A) or fluorescence (B) optics. Only one of the bipolar Schwann cells is labelled. Note that the other bipolar cell (putative Schwann cell) and the fibroblastic cells are unlabelled

phenotype) usually had a process-bearing morphology[31]. In human optic nerve cultures we also detected GFAP⁺ A2B5⁻ and GFAP⁺ A2B5⁺ astrocytes[2] (Figure 8.5). Although these generally appeared to correspond to the morphological categories of protoplasmic and fibrous astrocytes, respectively, this division was by no means distinct.

Effect of tissue culture medium on glial cell development in human fetal optic nerve cultures

It had been previously found in cultures of neonatal rat optic nerve that the A2B5⁺ progenitor cell could be induced to develop into a Type 2 astrocyte if grown in medium containing 10% FCS (DMEM–FCS), or a GC⁺

oligodendrocyte if grown in supplemented serum-free medium (DMEM–BS)[32,33]. The proportions of Type 1 astrocytes were not significantly different in cultures grown in the two different media. It was then established unequivocally that both Type 2 astrocytes and oligodendrocytes were actually derived from the one common precursor A2B5$^+$ GFAP$^-$ GC$^-$ cell. This represented the first example of glial cell developmental plasticity in the vertebrate CNS.

The question which we wished to ask was whether a similar developmental plasticity could be demonstrated in cultures of the human CNS. For this purpose we used 16–18-week-old human fetal optic nerve cultures and utilized the same strategies which had been used by Raff et al.[31,32]. The cultures were prepared as described above, and were examined at 1 and 3 days after plating. Cultures were grown in either DMEM–FCS or DMEM–BS. The results of these experiments are summarized in Table 8.1 in which the data obtained in rodent cultures are also shown.

It can be seen that at day 1 there was little difference between the composition of cultures grown in the two different media. The three main points of interest are:

(1)   60% of the cultures consisted of A2B5$^+$ cells;

(2)   there were already more GC$^+$ cells in cultures grown in DMEM–BS compared with DMEM–FCS;

(3)   there was a very small number of GC$^+$A2B5$^+$ cells in cultures grown in DMEM–BS but not DMEM–FCS.

Table 8.1 Proportion of marker-identified cell types in human optic nerve cultures

	Proportion (%) of total cells with the antigenic phenotype of:						
A2B5$^+$	A2B5$^+$ GFAP$^-$	A2B5$^+$ GFAP$^+$	A2B5$^-$ GFAP$^+$	GC$^+$ A2B5$^+$	GC$^+$ A2B5$^-$	GC$^+$ A2B5$^+$
Day 1:						
DMEM–BS	60 ± 2 (46 ± 4)	55 ± 3	5 ± 0.5 (0.6 ± 0.1)	6 ± 1	1 ± 0.3	2 ± 0.3 (11 ± 0.3)	57 ± 3
DMEM–FCS	61 ± 9 (44 ± 2)	54 ± 13	9 ± 3 (1.4 ± 0.1)	4 ± 2	0	0.7 ± 1 (8.5 ± 0.2)	55 ± 4
Day 3:						
DMEM–BS	59 ± 9 (30 ± 2)	47 ± 4	13 ± 5 (2.6 ± 0.1)	9 ± 7	0	8 ± 4 (27 ± 2)	48 ± 2
DMEM–FCS	56 ± 4 (13 ± 1)	26 ± 2	30 ± 4 (12.5 ± 1)	12 ± 4	0	1.1 ± 1.5 (0.9 ± 0.1)	50 ± 4

Numbers of marker-identified cell types are expressed as percentages of total cells in culture and are means ± SD of at least four separate experiments for Day 1 cultures and six separate experiments for Day 3 cultures. At least 200 cells/coverslip were counted in each experiment. Note that the decline in the A2B5$^+$ GFAP$^-$ and A2B5$^+$GC$^-$ populations correlates almost exactly with the increases in the A2B5$^+$ GFAP$^+$ and GC$^+$ populations of cells respectively. The total number of cells/coverslip, however, changed very little from Day 1 to Day 3. Numbers in parentheses refer to corresponding values in rat cultures derived from the data of Raff et al.[32] (Reprinted with permission from *Brain*, Vol. 109, No. 6, 1986, pp. 1261–77)

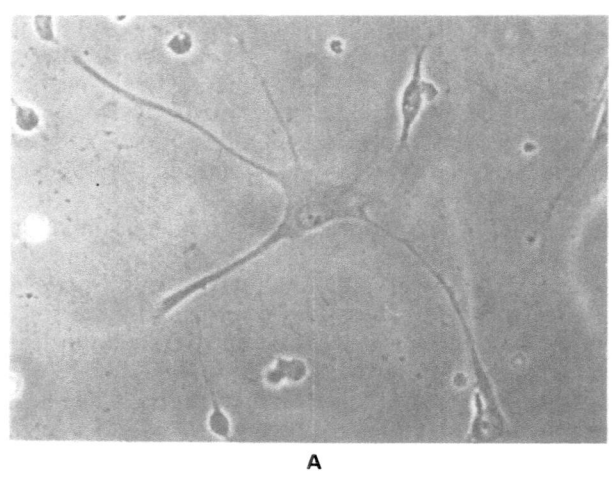

A

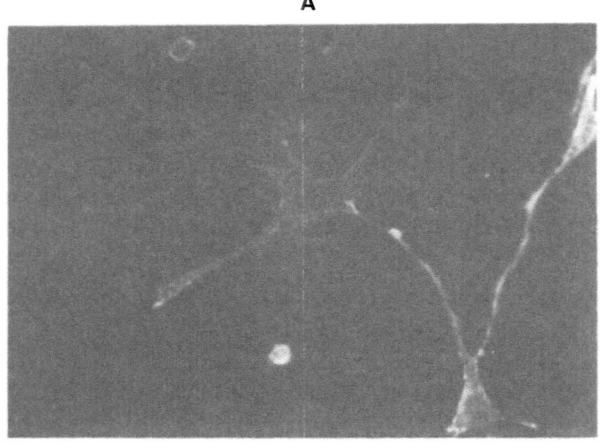

B

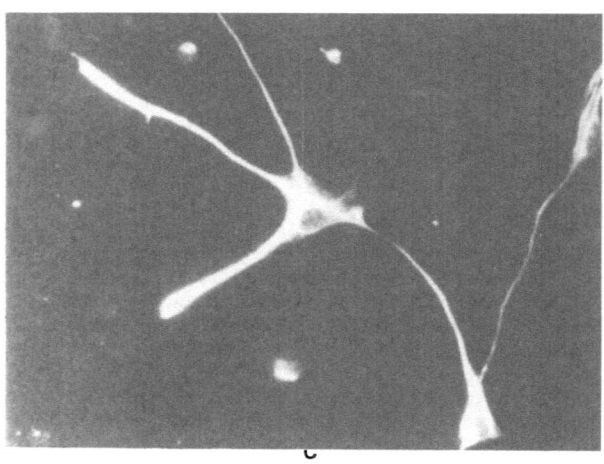

C

After 3 days in culture the proportion of the cell types showed marked differences from those seen at day 1, and many similarities to the corresponding rodent data. The main points of interest are:

(1) in cultures grown in DMEM–BS the percentage of GC^+ cells increased to a far greater extent than that observed in cultures grown in DMEM–FCS, but not as much as had been observed in rodent cultures. $GC^+ A2B5^+$ cells were no longer seen at this time;

(2) in cultures grown in DMEM–FCS the percentage of $GFAP^+ A2B5^+$ cells increased markedly with a corresponding decrease in the percentage of $A2B5^+ GFAP^-$ cells. The increase in $GFAP^+ A2B5^+$ cells seen in cultures grown in DMEM–BS was far less marked, although no increase at all had been observed in rodent optic nerve cultures grown in serum-free medium;

(3) the percentage of $GFAP^+ A2B5^-$ cells showed little difference in the two different media, as had been observed in rodent cultures.

These experiments showed remarkable similarities between the developmental behaviour of cultured human optic nerve cultures and the corresponding rodent cells, although there were several differences. It was clear that GC^+ oligodendrocyte development was favoured in serum-free medium and $GFAP^+ A2B5^+$ cell development promoted in DMEM–FCS, but it was not certain whether either or both of these cell types were actually derived from $A2B5^+$ precursor cells. We therefore treated freshly dissociated human optic nerve cells in suspension with A2B5 antibody and complement in order to eliminate the $A2B5^+$ cells in these initial cultures. It was found that no GC^+ cells developed in such treated cultures subsequently grown in DMEM–BS, showing that the development of this cell type required the presence of $A2B5^+$ cells. However, the proportion of $GFAP^+ A2B5^+$ cells in complement and A2B5 treated cultures and subsequently grown in DMEM–FCS was only reduced to 50% of untreated control cultures. This contrasts with the finding in rodent cultures in which no Type 2 astrocytes developed after treatment with A2B5 and complement. Nevertheless, at least some human $GFAP^+ A2B5^+$ cells appear to require the presence of $A2B5^+$ cells for their development in culture. In both human and rodent cultures the proportion of $GFAP^+ A2B5^-$ cells was not affected by such prior treatment, indicating that these cells are not derived from $A2B5^+$ precursor cells. Unfortunately it did not prove consistently possible to pulse label human cultures with A2B5 in order to show unequivocally that GC^+ and $GFAP^+ A2B5^+$ cells were derived from $A2B5^+$ precursor cells, as had been possible in rodent cultures.

Figure 8.5 $GFAP^+ A2B5^+$ astrocytes. After 3 days cells in human optic nerve culture were incubated with monoclonal A2B5 antibody followed by goat anti-mouse Ig conjugated to rhodamine. The cells were then fixed in methanol and labelled with rabbit anti-GFAP serum followed by sheep anti-rabbit Ig conjugated to fluorescein. Cultures were then viewed with phase contrast (A), rhodamine (B) and fluorescein (C) optics. Three cells in the field are double-labelled with both antibodies

It should be appreciated that the gestational age at which we studied the human cells did not correspond to the stage of development in neonatal rat optic nerves. It is therefore not surprising that the findings in the two culture systems showed differences. Nevertheless these studies on human glial cells do show several striking similarities to those previously performed in rodent glial cell cultures and demonstrated a degree of developmental plasticity in the human CNS. Despite the severe scarcity of such human tissues we hope to pursue these lines of approach in the future and to continue to use such culture systems in clinical studies.

EXPERIMENTAL INFECTION OF HUMAN NEURAL CELL CULTURES WITH HERPES SIMPLEX VIRUS

One way in which these human neural cell cultures can be exploited is by examining the susceptibility of defined human glial and neuronal cell types to infection with viruses. We chose herpes simplex virus (HSV) in these studies for a variety of reasons. HSV is a significant cause of human neurological diseases, in particular HSV encephalitis. Although this disease is rare, it is still the most frequent fatal sporadic acute encephalitis occurring in humans, and its pathogenesis is not well understood[34]. HSV is also associated with aseptic meningitis[35], ascending myelitis[36], radiculomyelopathy[37] and congenital malformations of the CNS[38]. Moreover, the ability of HSV to produce latent infections in neurons in the sensory ganglia of humans and experimental animals is well known[39]. Human neural cell cultures afford an opportunity to construct *in vitro* models of HSV latency which may have more relevance to clinical disease than similar *in vitro* studies using cultures derived from animal tissues. The studies summarized below represent the first attempt to combine virological and cell marker methodologies to define the *in vitro* susceptibility of human neural cells to infection with HSV[17]. These experiments were performed after the initial human cell marker techniques were established but before the recent work on glial progenitor cells was carried out. Future experiments will focus on infection of glial precursor cells and different astrocyte populations.

The techniques used in these studies have been given in detail elsewhere[1,2]. In brief, dissociated primary neural cultures derived from human fetal DRG, spinal cord and brain were prepared as outlined above. Serially passaged dissociated cultures were also prepared from primary brain cultures. Both wild-type virus (ts^+syn^+) and a range of temperature-sensitive (*ts*) mutant viruses ($ts\,syn$ or $ts\,syn^+$) were used[40,41] (Table 8.2). A *ts* mutant has a mutation which renders it unable to produce infectious virus progeny at the non-permissive temperature (38.5°C) but it can still replicate normally at the permissive temperature (31°C), whereas wild-type virus can replicate at both 31°C and 38.5°C[42,43]. The terms *syn* and *syn*$^+$ denote syncytial and nonsyncytial plaque morphologies, respectively. HSV infection of cultures growing on glass coverslips was carried out initially using a range of multiplicities of infection (MOI), but in the majority of experiments an MOI of ten plaque-forming units (PFU) per cell was used. The cells were examined at 6 and 24 h after infection for the development of a cytopathic effect, and expression of cell markers and viral

Table 8.2 Viruses used to infect human neural cell cultures

Wild-type	HSV-1 (ts^+syn^+) Glasgow Strain 17
Wild-type	HSV-2 (HG52)
Mutants	HSV-1 (thymidine kinase deficient ts^+tk^{-7})
	HSV-1 temperature-sensitive: tsGsyn^+
	tsHsyn^+
	tsIsyn
	tsKsyn^+
	tsLsyn^+
	tsSsyn^+
	HSV-2 temperature-sensitive: ts5

antigens using a variety of polyclonal and monoclonal antibodies in indirect immunofluorescence assays. Both single-labelling and double-labelling experiments were carried out to visualize viral antigen expression on marker-identified cell types. BHK21/C13 cell cultures, which are susceptible to HSV infection, were also infected in control experiments.

Infection of primary human cell cultures

Primary cultures of DRG contained approximately 30% A2B5$^+$ neurons, 30% Schwann cells and 40% FN$^+$ fibroblasts. Brain and spinal cord cultures comprised approximately 50% and 10% neurons, 30% and 35% GFAP$^+$ astrocytes, and 10% and 60% fibroblasts, respectively. GC$^+$ cells were not identifiable with markers in these cultures because of the younger age (15–16-week-old) of the fetuses used in these studies. The viruses studied in these cultures included the wild-type ts^+syn^+ and the mutants tsGsyn^+ and tsKsyn^+. The detailed findings will not be given here as they have been published previously[7]. The essential points which emerged were:

(1) the percentage of neurons showing a CPE (recognized by rounding up of cells with loss of usual cell morphology) at all time points was lower compared with that of astrocytes, fibroblasts and Schwann cells;

(2) irrespective of the virus used or the source of the cultured cells the proportions of central and peripheral neurons expressing viral antigens was markedly lower than that observed with the other cell types.

These observations were apparent at both 6 and 24 h after infection. In the few experiments in which higher MOI were used or the cultures were examined after 48 h the same patterns of cytopathic effect (CPE) and antigen expression were observed. The ts mutants differed considerably in their effects in these cultures. For example, tsKsyn^+ induced very little CPE and weak viral antigen expression at both the permissive and non-permissive temperatures, which is consistent with the processing defect of this mutant reported in neuroblastoma cells[44]. By contrast tsGsyn^+ was far more cytopathic in human cells. It should be pointed out that failure of an infected cell to express large amounts of viral antigen may result not only from the cell's greater resistance to infection with HSV but also from suppression and/or delay of antigen expression due possibly to a slower lytic growth cycle. Moreover, we believe that at least some astrocytes lose GFAP during HSV infection[7]. These observations of HSV-1 infection in neurons are entirely consistent with earlier studies performed without cell markers[45]

Infection of passaged human cell cultures

Passaged cultures comprised 15–30% GFAP$^+$ astrocytes, 60–70% GFAP$^-$ FN$^+$ cells and 10–20% GFAP$^-$ FN$^-$ cells. These cultures were studied at passage numbers 1–5, and the effects of all the viruses shown in Table 7.2 were tested[7]. One-step growth curve experiments showed that HSV was actually replicating in infected cultures (Figure 8.6).

Two general patterns of infection clearly emerged in these lytically infected passaged cell cultures:

(1)   the mutants ts^+tk^{-7}, tsHsyn^+, tsIsyn and tsSsyn^+ did not induce either CPE or viral antigen expression at 6 h post-infection, but at 24 h a CPE was observed in 50–80% of cells and weak antigen expression was seen in the majority of cells (Figure 8.7);

(2)   wild-type HSV-1 and HSV-2, tsGsyn^+ and tsLsyn^+ produced a CPE in 30% of the cells and weak viral antigen expression in the majority of cells at 6 h post-infection, and a marked CPE with a high level of antigen expression in most cells at 24 h.

Of all the viruses studied, tsGsyn^+ was the most cytopathic for human and BHK21/C13 cells, probably due at least in part to its low particle/PFU ratio[7,43], and tsKsyn^+ was the least cytopathic. The biological significance of these heterogeneous patterns of infection with these mutants is uncertain at present.

Relevance of *in vitro* studies to natural HSV infection

Although it can be argued that experimental viral infection of human cell cultures is more meaningful in terms of human disease than similar infection of cultured animal neural cells, the question obviously arises as to how such *in vitro* observations can be interpreted in relation to naturally occurring *in vivo* infection. Experimental *in vivo* infection is obviously not possible in humans. Our observations that HSV-1 and HSV-2 did not show different behaviour in these cultures was surprising since clinical evidence suggests that HSV-1 is more neurovirulent than HSV-2. For example, HSV-1 is a much more frequent cause of HSV encephalitis than HSV-2, the latter virus being particularly associated with encephalitis in neonates[38] and immunocompromised individuals including those with the acquired immunodeficiency syndrome (AIDS)[46]. However, our studies of HSV-2 were limited to passaged cultures and a more extensive analysis of its activity in primary cultures would be required before more definite conclusions can be reached on the comparative *in vitro* neurovirulence of these two viruses.

Although we demonstrated HSV infection of both cultured neurons and glial cells, the infection was far more restricted in neurons in terms of both CPE and the development of viral antigen expression. Since the cell cultures were heterogeneous the amounts of infectious virus released by specific neuronal populations is not known. How do these findings relate to clinical disease? In HSV encephalitis of humans, viral antigens have been located using morphological criteria in both neurons and glial cells[47]. However, double-labelling experiments in which viral antigens are localized in marker-identified cell types in these cases have not been reported as yet. It would be

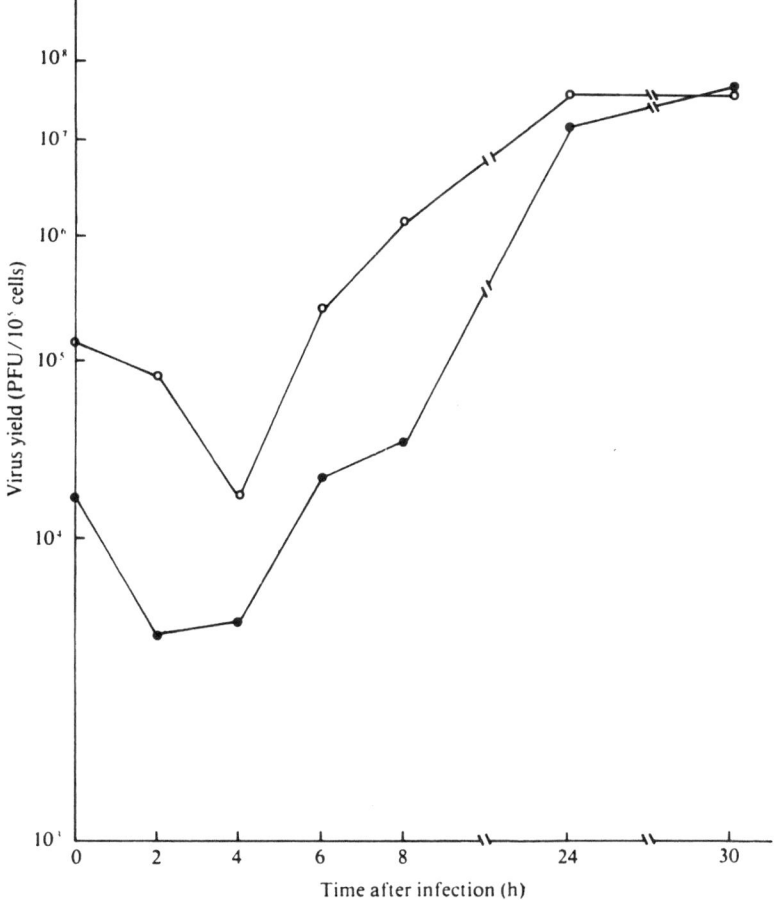

Figure 8.6   One-step growth curves of HSV-1 in BHK21/C13 and fetal human brain cells. One-step growth curves of HSV-1 wild type virus (ts^+syn^+) in fetal human brain cells (open circles) and BHK21/C13 cells (filled circles). Cells were infected at a multiplicity of infection of 2 PFU/cell and incubated at 37°C. At various times post-infection, cultures were harvested and the virus released from the cells by sonication. Virus yields were titrated on BHK21/C13 cells at 37°C. (Reprinted, with permission, from *Brain* Vol. 106, pp. 101–19, 1980)

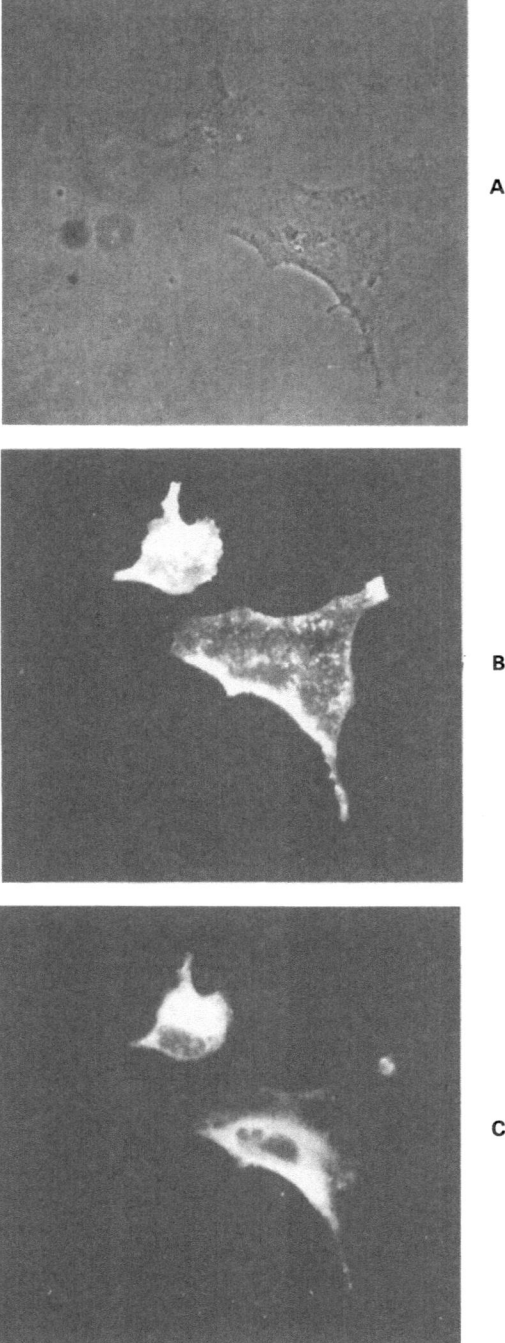

particularly helpful to know the very early cell tropism of HSV in the disease, and experimental HSV infection in animals should be useful in determining the clinical relevance of our *in vitro* findings.

With regard to HSV latency there is abundant evidence to show that the neuron is the cell which harbours latent virus[48,49]. How is it possible to reconcile this fact with the observation that neurons appear to be more resistant than other cell types to acute lytic infection with HSV? One possibility is that a low rate of virus production in the more resistant neurons might help to maintain HSV infection in sensory ganglia without significant progression[50]. Moreover, experimental HSV latency in cultured human DRG neurons following acute infection has been demonstrated by Wigdahl *et al.*[51] Heat-shock proteins, which are up-regulated in cultured human DRG neurons following HSV infection[52], may also play a role in the establishment and regulation of latency and it is of interest that HSV infection followed by heat stress has been used to establish a latent infection in these cells[51]. However, a more detailed evaluation of the clinical relevance of the *in vitro* findings described here must await further studies on human and animal pathological material.

Acknowledgements

It is a pleasure to acknowledge the collaboration of the following colleagues in the studies described here: Drs M. C. Raff, R. P. Lisak, G. B. Clements, S. M. Brown, J. Fok-Seang and M. Noble. Much of the work described was supported by the National Fund for Research into Crippling Diseases, and the British Multiple Sclerosis Society. Permission to work with human fetal tissues was granted by the ethical committees of University College, London, University College Hospital, Inverclyde Health District and the Institute of Neurology, London.

References

1. Kennedy, P. G. E., Lisak, R. P. and Raff, M. C. (1980). Cell type-specific markers for human glial and neuronal cells in culture. *Lab. Invest.*, **43**, 342–51
2. Kennedy, P. G. E. and Fok-Seang, J. (1986). Studies on the development, antigenic phenotype and function of human glial cells in tissue culture. *Brain,* **109**, 1261–77
3. Kennedy, P. G. E. and Lisak, R. P. (1979). A search for antibodies against glial cells in the serum and cerebrospinal fluid of patients with multiple sclerosis and Guillain-Barré syndrome. *J. Neurol. Sci.*, **44**, 125–33
4. Kennedy, P. G. E. and Lisak, R. P. (1981). Do patients with demyelinating diseases have antibodies against human glial cells in their sera? *J. Neurol. Neurosurg. Psychiatr.*, **44**, 164–7

Figure 8.7 HSV$^+$ GFAP$^+$ cells in infected passaged human brain culture. 24 h after infection with 10 PFU/cell of the HSV-1 mutant *tsHsyn*$^+$, cells in passaged human fetal brain cultures were labelled with polyclonal anti-HSV serum followed by sheep anti-rabbit Ig conjugated to fluorescein and fixed in acid–alcohol. After washing they were incubated with rabbit anti-GFAP serum followed by goat anti-rabbit Ig conjugated to rhodamine and viewed with phase contrast (A), fluorescein (B) or rhodamine (C) optics. The two cells show surface labelling with rabbit anti-HSV serum and intracellular labelling with anti-GFAP serum

5. Watts, H., Kennedy, P. G. E. and Thomas, M. (1981). The significance of antineuronal antibodies in Alzheimer's disease. *J. Neuroimmunol.*, **1**, 107-16
6. Wroblewska, Z., Kennedy, P. G. E., Wellish, M., Lisak, R. P. and Gilden, D. H. (1982). Demonstration of JC virus by immunofluorescence in multiple cell types in experimentally infected adult human brain cell cultures. *J. Neurol. Sci.*, **54**, 189-96
7. Kennedy, P. G. E., Clements, G. B. and Brown, S. M. (1983). Differential susceptibility of human neural cell types in culture to infection with herpes simplex virus. *Brain*, **106**, 101-19
8. Raff, M. C., Mirsky, R., Fields, K. L., Lisak, R. P., Dorfman, S. H., Silberberg, D. H., Gregson, N. A., Leibowitz, S. and Kennedy, M. C. (1978). Galactocerebroside is a specific cell-surface antigenic marker for oligodendrocytes in culture. *Nature*, **274**, 813-6
9. Raff, M. C., Fields, K. L., Hakomori, S., Mirsky, R., Pruss, R. M. and Winter, J. (1979). Cell-type specific markers for distinguishing and studying neurons and the major classes of glial cells in culture. *Brain Res.*, **174**, 283-308
10. Dickson, J. G., Flanigan, T. P. and Walsh, F. S. (1982). Cell-surface antigens of human fetal brain and dorsal root ganglion cells in tissue culture. In Rowland, L. P. (ed.). *Human Motor Neuron Diseases*. pp. 435-51. (New York: Raven Press)
11. Kennedy, P. G. E. (1982). Neural cell markers and their applications to neurology. *J. Neuroimmunol.*, **2**, 35-53
12. Eng. L. P., Vanderhaeghen, J. J., Bignami, A. and Gerstl, B. (1971). An acidic protein isolated from fibrous astocytes. *Brain Res.*, **28**, 351-4
13. Dahl, D. and Bignami, A. (1973). Immunochemical and immunofluorescence studies of the glial fibrillary acidic protein in vertebrates. *Brain Res.*, **61**, 279-93
14. Schachner, M., Hedley-Whyte, E. T., Hsu, D. W., Schoonmaker, G. and Bignami, A. (1977). Ultrastructural localization of glial fibrillary acidic protein in mouse cerebellum by immunperoxidase labelling. *J. Cell. Biol.*, **75**, 67-73
15. Cotmore, S. F., Crowhurst, S. A. and Waterfield, M. D. (1981). Purification of Thy-1 related glycoproteins from human brain and fibroblasts: comparisons between these molecules and murine glycoproteins carrying Thy-1 and Thy-1,2. *Eur. J. Biochem.*, **11**, 597-603
16. Gilden, D. H., Wroblewska, Z., Eng, L. F. and Rorke, L. B. (1976). Human brain in tissue culture, part 5 (identification of glial cells by immunofluorescence). *J. Neurol. Sci.*, **29**, 177-84
17. Bunge, R. P. (1968). Glial cells and the central myelin sheath. *Physiol. Rev.*, **48**, 197-251
18. Norton, W. T. and Autilio, L. A. (1966). The lipid composition of purified bovine brain myelin. *J. Neurochem.*, **13**, 213-22
19. Lisak, R. P., Pleasure, D. E., Silberberg, D. H., Manning, M. C. and Saida, T. (1981). Long-term culture of bovine oligodendrocytes isolated with a Percoll gradient. *Brain Res.*, **223**, 107-22
20. Kim, S. U., Sato, Y., Silberberg, D. H., Pleasure, D. E. and Rorke, L. B. (1983). Long-term culture of human oligodendrocytes. Isolation, growth and identification. *J. Neurol. Sci.*, **62**, 295-301
21. Dimpfel, W., Huang, R. T. C. and Habermann, E. (1977). Gangliosides in nervous tissue cultures and binding of ^{125}I-labelled tetanus toxin – A neuronal marker. *J. Neurochem.*, **29**, 329-34
22. Mirsky, R., Wendon, L. M. B., Black, P., Stolkin, C. and Bray, D. (1978). Tetanus toxin – A cell surface marker for neurones in culture. *Brain Res.*, **148**, 251-9
23. Cohen, J. and Selvendran, S. (1981). A neuronal cell-surface antigen is found in the CNS but not in peripheral neurones. *Nature*, **291**, 421-3
24. Vulliamy, T., Rattray, S. and Mirsky, R. (1981). Cell-surface antigen distinguishes sensory and autonomic peripheral neurones from central neurones. *Nature*, **291**, 418-20
25. Anderton, B. H., Thorpe, R., Cohen, J., Selvendran, S. and Woodhams, P. (1980). Specific localization by immunofluorescence of 10 nm filament polypeptides. *J. Neurocytol.*, **9**, 835-44
26. Wood, J. N. and Anderton, B. (1981). Monoclonal antibodies to mammalian neurofilaments. *Biosci. Rep.*, **1**, 263-8
27. Eisenbarth, G. S., Walsh, F. S. and Nirenberg, M. (1979). Monoclonal antibody to a plasma membrane antigen of neurones. *Proc. Natl. Acad. Sci. USA*, **76**, 4913-7

28. Fields, K. L., Gosling, C., Megson, M. and Stern, P. L. (1975). New cell surface antigens in rat defined by tumours of the nervous system. *Proc. Natl. Acad. Sci. USA*, **72**, 1296-300
29. Brockes, J. P., Fields, K. L. and Raff, M. C. (1977). A surface antigenic marker for rat Schwann cells. *Nature*, **266**, 364-6
30. Vaheri, A., Ruoslahti, E., Westermark, B. and Ponten, J. (1976). A common cell-type-specific surface antigen in cultured human glial cells and fibroblasts; loss in malignant cells. *J. Exp. Med.*, **143**, 64-72
31. Raff, M. C., Abney, E. R., Cohen, J., Lindsay, R. and Noble, M. (1983). Two types of astrocytes in cultures of developing rat white matter: differences in morphology, surface gangliosides and growth characteristics. *J. Neurosci.*, **3**, 1289-300
32. Raff, M. C., Miller, R. H. and Noble, M. (1983). A glial progenitor cell that develops *in vitro* into an astrocyte or an oligodendrocyte depending on culture medium. *Nature*, **303**, 390-6
33. Bottenstein, J. E. and Sato, G. H. (1979). Growth of a rat neuroblastoma cell line in serum-free supplemented medium. *Proc. Natl. Acad. Sci. USA*, **76**, 514-7
34. Kennedy, P. G. E. (1984). Herpes simplex virus and the nervous system. *Postgrad. Med. J.*, **60**, 253-9
35. Craig, C. P. and Nahmias, A. J. (1973). Different patterns of neurologic involvement with herpes simplex virus type 1 and 2: Isolation of herpes simplex virus type 2 from the buffy coat of two adults with meningitis. *J. Infect. Dis.*, **127**, 365-72
36. Klastersky, J., Cappel, R., Snoeck, J. M., Flament, J. and Thiry, L. (1972). Ascending myelitis in association with herpes simplex virus. *N. Engl. J. Med.*, **287**, 182-4
37. Caplan, L. R., Kleeman, F. J. and Berg, S. (1977). Urinary retention probably secondary to herpes genitalis. *N. Engl. J. Med.*, **279**, 920-1
38. Johnson, R. T. (1982). *Viral Infections of the Nervous System*. (New York: Raven Press)
39. Bartinger, J. R. (1975). Herpes simplex virus infection of nervous tissue in animals and man. *Progr. Med. Virol.*, **20**, 1-26
40. Timbury, M. C. (1971). Temperature-sensitive mutants of herpes simplex virus type 2. *J. Gen. Virol.*, **13**, 373-6
41. Brown, S. M., Ritchie, D. A. and Subak-Sharpe, J. H. (1973). Genetic studies with herpes simplex virus type 1. The isolation of temperature sensitive mutants, their arrangement into complementation groups and recombination analysis leading to a linkage map. *J. Gen. Virol.*, **18**, 329-46
42. Clements, G. B. (1975). Selection of biochemically variant, in some cases mutant, mammalian cells in culture. In Klein, G. and Weinhouse, S. (eds.) *Advances in Cancer Research*. pp. 274-380. (New York and London: Academic Press)
43. Subak-Sharpe, J. H. (1973). The genetics of herpes virus. *Cancer Res.*, **32**, 1385-92
44. Gerdes, J. C., Marsden, H. S., Cook, M. L. and Stevens, J. G. (1979). Acute infection of differentiated neuroblastoma cells by latency-positive and latency-negative HSV *ts* mutants. *Virology*, **94**, 430-41
45. Rajcani, J. and Scott, B. S. (1972). Growth of herpes simplex virus in cultures of dissociated human nervous tissue. *Acta Virol.*, **16**, 25-30
46. Dix, R. D., Waitzman, D. M., Follansbee, S., Pearson, B. S., Mendelson, T., Smith, P., Davis, R. L. and Mills, J. (1985). Herpes simplex virus type 2 encephalitis in two homosexual men with persistent lymphadenopathy. *Ann. Neurol.*, **17**, 203-6
47. Booss, J. and Esiri, M. M. (1986). *Viral Encephalitis. Pathology, Diagnosis and Management*. (Oxford: Blackwell Scientific Publications)
48. McLennan, J. L. and Darby, G. C. (1980). Herpes simplex virus latency. The cellular location of virus in dorsal root ganglia and the fate of the infected cell following virus activation. *J. Gen. Virol.*, **51**, 233-43
49. Kennedy, P. G. E., Al-Saadi, S. A. and Clements, G. B. (1983). Reactivation of latent herpes simplex virus from dissociated identified dorsal root ganglion cells in culture. *J. Gen. Virol.*, **64**, 1629-35
50. Vahlne, A., Nystrom, B., Sandberg, M., Hamberger, A. and Lycke, E. (1978). Attachment of herpes simplex virus to neurones and glial cells. *J. Gen. Virol.*, **44**, 359-71
51. Wigdahl, B., Smith, C. A., Traglia, H. M. and Rapp, F. (1984). Herpes simplex virus latency in isolated human neurones. *Proc. Natl. Acad. Sci. USA*, **81**, 6217-21

52. Kennedy, P. G. E., LaThangue, N. B., Chan, W. L. and Clements, G. B. (1985). Cultured human neural cells accumulate a heat-shock protein during acute herpes simplex virus infection. *Neurosci. Lett.*, **61**, 321–6

9
Human Cytomegalovirus: The Major Envelope Glycoprotein as a Candidate for a Subunit Vaccine

M. MACH, U. UTZ AND B. FLECKENSTEIN

INTRODUCTION

Human cytomegalovirus (HCMV) is a member of the herpesvirus family and, like other human herpesviruses, is ubiquitous in human populations. The virus has been associated with a variety of clinical syndromes. HCMV is the most common cause of congenital virus infection in newborns and it accounts for a significant number of developmentally disabled children. The best known and most devastating manifestation of HCMV infection is cytomegalic inclusion disease, characterized by hepatosplenomegaly, jaundice, thrombocytopenic purpura, low birth weight, microcephaly and damage to the central nervous system. The syndrome has an incidence of at least one in 3000 live births in the United States[1]. In addition, high incidence of HCMV infection, most often occurring as interstitial pneumonia or an infectious mononucleosis-like syndrome, has been reported after bone marrow transplants[2] as well as in immunosuppressed patients with renal or cardiac transplants[3]. With the increased incidence of organ transplantation and the concomitant use of potent immunosuppressive agents, HCMV will continue to be an important risk factor in these populations.

Many aspects of the immune response to HCMV remain poorly understood. However, both antiviral antibodies and cellular functions have been shown to be important in controlling HCMV infections. Recent findings have provided evidence that the envelope glycoproteins are immunogenic in humans and can elicit neutralizing antibodies[4-6]. Sera from patients convalescing from HCMV infections can neutralize infectious virus in tissue culture and contain antibodies that are reactive with glycoproteins. In addition, monoclonal antibodies have been isolated which recognize a major viral glycoprotein of about 55–58 kDa in size (gp58) and also neutralize infectious virus[6,7]. Since a generally applicable HCMV vaccine can probably not be developed by classical virological procedures, attempts should be based on the expression cloning of viral glycoprotein genes. Here we describe the identification and structural analysis of the gene encoding the major glycoprotein of human cytomegalovirus.

GENERATION AND CHARACTERIZATION OF A MONOSPECIFIC ANTISERUM AGAINST THE MAJOR GLYCOPROTEIN (gp58)

Human cytomegalovirus contains approximately 25–35 structural proteins (Figure 9.1) of which three to eight are glycosylated. In polyacrylamide gel analysis the [³H]glucosamine-labelled virion glycoproteins show a complex electrophoretic profile composed of at least two well-labelled bands and several minor ones[8,9]. This heterogeneity makes it difficult to determine the exact number and to assign accurate molecular weights of virion glycoproteins. One major glycosylated component of purified virions is a polypeptide of approximately 58 kDa in size. Monoclonal antibodies against this protein have been isolated in several laboratories[4-7]. The mature protein is derived from a larger, probably non-glycosylated, precursor molecule of about 95–115 kDa in size via proteolytic cleavage. The molecule

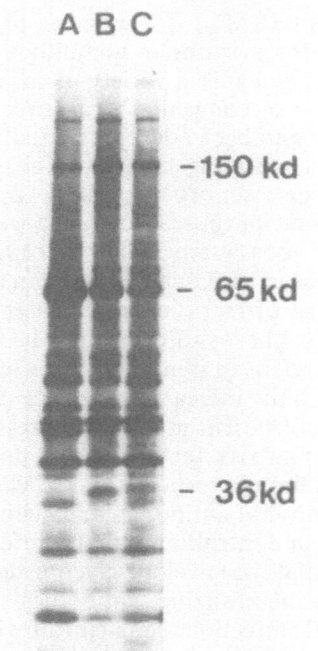

A B C

−150 kd

− 65 kd

− 36 kd

Figure 9.1 Structural proteins of HCMV particles. Human foreskin fibroblasts were infected with HCMV strain AD169 and labelled metabolically with [³⁵S]methionine from 90–120 h post infection. Extracellular particles were separated on a glycerol-tartrate gradient[23] and subjected to SDS–polyacrylamide gel electrophoresis. Lanes: A, dense bodies; B, virions; C, NIEPs (non-infectious enveloped particles). Bands identified: 150, 65 and 36 kDA

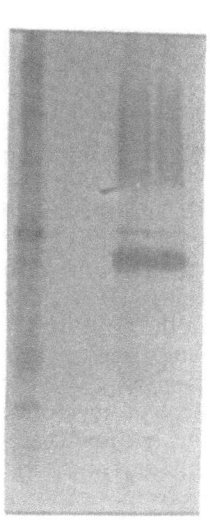
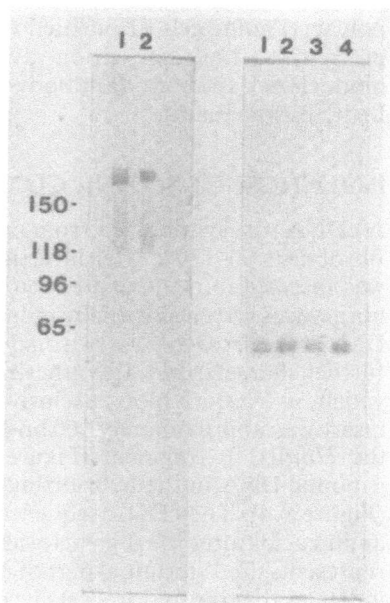

Figure 9.2 Immunological characterization of the monospecific gp58 serum with bands identified by kDa size. (A; left) [^{35}S]Methionine-labelled AD169 virions were purified and disrupted in 0.1% NP40–PBS. Following immune precipitation with HCMV non-immune rabbit serum (lane 0), gp58 serum (lane a 58) and a mouse monoclonal antibody (lane 7-17) the precipitates were eluted in SDS and β-mercaptoethanol-containing buffer and analysed on 10% SDS–PAGE. Lane Vir contains purified, disrupted, non-immune precipitated virions. (B; right) Western blot analyses of HCMV structural proteins. Proteins from gradient-purified HCMV AD169 virions were separated on a 10% SDS–PAGE in the presence (right panel) or absence (left panel) of β-mercaptoethanol. After transfer to nitrocellulose the membrane was incubated with gp58 serum followed by horseradish peroxidase labelled Protein-A, and 4-chloro-1-naphthol as the chromogen. Lanes: 1, extracellular HCMV particles; 2, gradient purified virions; 3, gradient-purified NIEPs; 4, gradient-purified dense bodies

is located within the outer envelope of the virus[9] and exists as a complex together with at least one other protein, held together by disulphide bonds. This can be demonstrated in Western blots and immunoprecipitations using monoclonal antibodies or monospecific antisera. The Western blot technique using a monospecific serum performed in the presence of reducing agents such as β-mercaptoethanol or dithiothreitol, allow recognition of a single polypeptide of 58 kDa in size. In the absence of reducing agents, the same serum reacts only with high molecular weight complexes (Figure 9.2a). These complexes most probably do not represent homopolymers of gp58, since in immunoprecipitations using either monoclonal antibodies or monospecific sera, high molecular weight polypeptides are precipitated even in the presence of reducing agents (Figure 9.2b). In addition, Britt and Auger[10] have recently shown by proteolytic peptide mapping that the high molecular weight polypeptides share no homology with gp58. We have generated a monospecific antiserum against gp58 in rabbits, using protein extracted from preparative polyacrylamide gels of purified virions. As shown in Figure 9.2, this serum precipitated the same virion glycoproteins as a well characterized mouse monoclonal antibody (antibody 7-17 was kindly provided by Dr William Britt, Birmingham).

ISOLATION OF A cDNA CLONE ENCODING gp58

A cDNA was synthesized from poly A^+ RNA isolated from human foreskin fibroblasts 96–120 h after infection with HCMV strain AD169. The cDNA was inserted into the expression vector λ-gt11[11], and the resulting cDNA-library was screened with the monospecific antiserum. One clone (designated UM-2) which reacted very strongly with the anti-gp58 serum was purified and further characterized. It synthesized a β-galactosidase-HCMV fusion protein which, in Western blots, exclusively reacted with the antiserum. The cDNA insert was approximately 400 bp in size and it hybridized to the right end of the *Hind*III F fragment (Figure 9.3)[12]. The inserted cDNA and HCMV genomic DNA in the hybridizing region had identical nucleotide sequences (Figure 9.4). The cDNA sequence contains a translational stop signal eight amino acids after the β-galactosidase–HCMV junction indicating that UM-2 represents the 3'-terminal part of the open reading frame encoding gp58. This indicates that the major viral glycoprotein gp58 originates from the carboxy-terminal part of the precursor molecule.

COMPARISON OF gp58 WITH GLYCOPROTEIN GENES OF HUMAN HERPES VIRUSES

HCMV has been the last human herpesvirus lacking information about envelope glycoprotein genes; with our data now available, a comparison of these genes is possible for the first time. The gp58 open reading frame of HCMV is located immediately upstream of the DNA polymerase gene[13]. This arrangement shows striking homologies with the other human herpesviruses. In Epstein-Barr virus (EBV) an open reading frame (BALF4) is also located immediately upstream of the DNA polymerase gene[14]. The translational product of this gene, however, seems not to be abundant in EBV virions. The immunodominant membrane antigens of EBV consist largely of two high molecular weight components of 350 kDa and 220 kDa in size[15]. Direction of transcription is the same for both genes. An identical gene arrangement is found in varicella zoster virus (VZV). The open reading frame coding for gpII, a major glycoprotein of VZV, is located close to the DNA polymerase gene[16]. The mature gpII polypeptide migrates as a slightly resolved doublet of ca 60 kDa in reducing SDS–PAGE but as a single 120–140 kDa species in non-reducing gels. The 60 kDa protein is derived from a larger precursor molecule via proteolytic cleavage and it exits within the viral envelope as a heterodimer held together by disulphide bands[17]. The herpes simplex virus (HSV) gene coding for glycoprotein B (gB) also maps in the vicinity of the DNA polymerase gene. In this case, however, the direction of transcription is opposite[18,19].

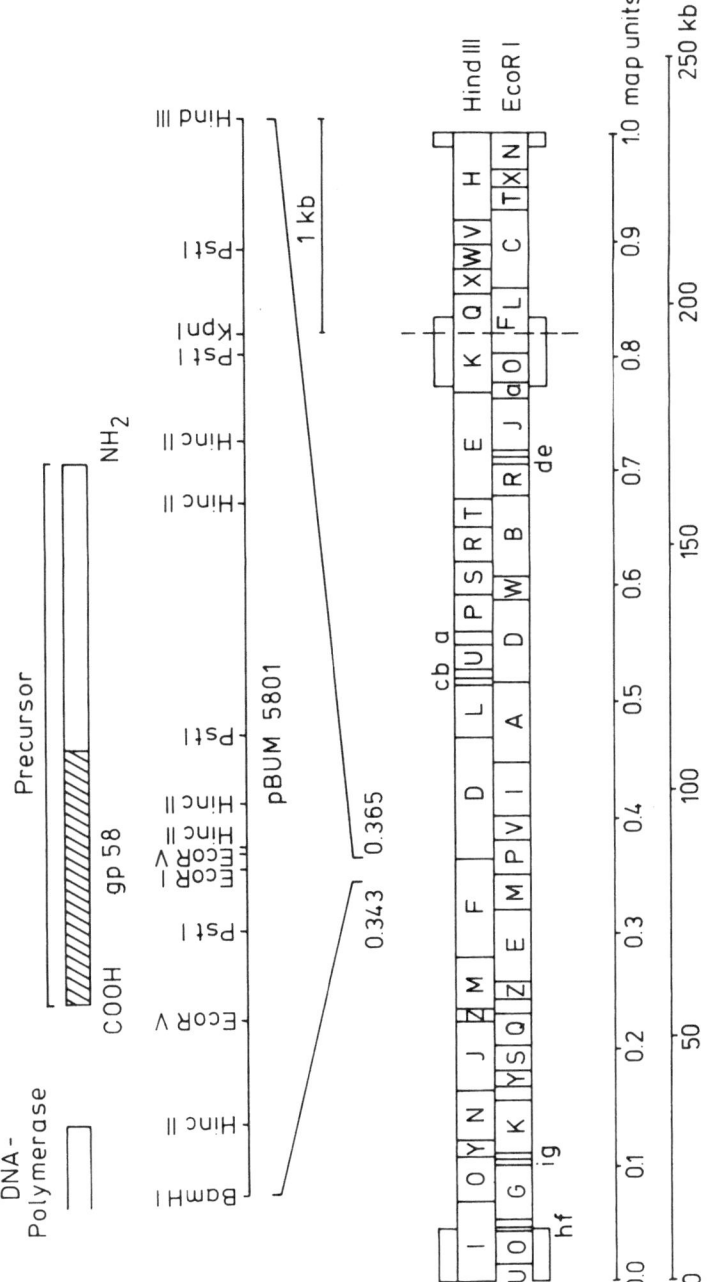

Figure 9.3 The coding region of gp58

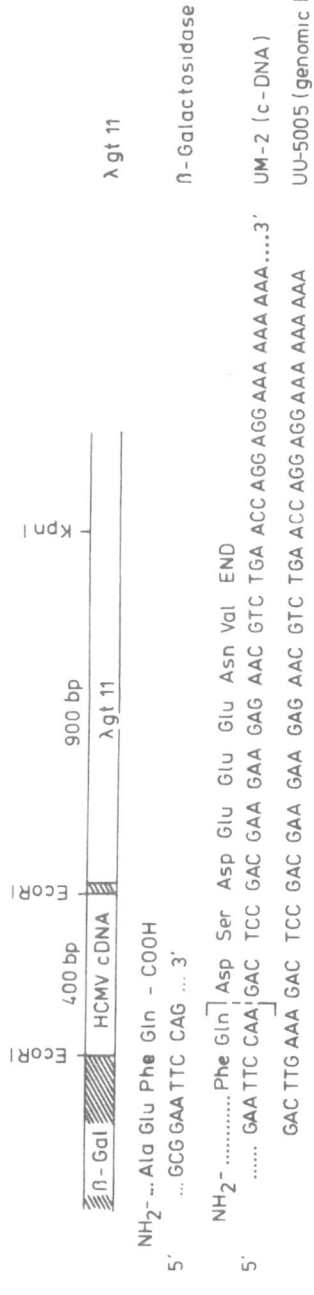

Figure 9.4 Sequence comparison of the gp58 cDNA and genomic HCMV DNA. The nucleotide and amino acid sequences surrounding the β-galactosidase–gp58 junction are shown. It should be noted that the GAATTCC sequence in the UM-2 construct is derived from the EcoRI linkers used in the original cDNA cloning procedure[12]

All the glycoprotein reading frames show significant homology at the amino acid level. It therefore appears as if a common genomic arrangement has been conserved in human herpesviruses regardless of their evolutionary distance.

Though the role of HCMV gp58 in virus infection has not yet been defined, its counterparts in HSV and VZV are major targets for antibody-mediated neutralization and have been implicated in the penetration of the virus into cells[20,17]. In addition, most monoclonal antibodies against gp58 are able to neutralize virus *in vitro* suggesting an important role of this protein in the infectious process.

CLINICAL ISOLATES

A necessary requirement for the potential use of gp58 as a subunit vaccine is the presence of this protein in most, if not all, HCMV isolates. HCMV, like other herpesviruses, shows a significant degree of antigenic diversity as well as differences at the genomic level[21,22]. In order to evaluate the variability of gp58 we have started to analyse clinical isolates for the presence of this polypeptide. Using Western blots with the monospecific antiserum, we have detected gp58 in all clinical isolates so far, supporting the notion that this protein may indeed be a candidate for the development of a subunit vaccine.

CONCLUDING REMARKS

This study describes the identification of the glycoprotein gene gp58 of human cytomegalovirus and the characterization of the respective polypeptide. It is a major target for neutralizing antibodies. Current efforts are aimed at the expression of neutralizing epitopes in bacterial and eukaryotic systems. However, it remains to be proved that humoral antibodies can protect the human organism against primary virus infection.

References

1. Onorato, I. M., Morens, D. M., Martone, W. J. and Stansfield, S. K. (1985). Epidemiology of cytomegalovirus infections: Recommendations for prevention and control. *Rev. Infect. Dis.*, **7**, 479-97
2. Meyers, J. D. (1985). Cytomegalovirus infection after organ allografting: Prospects for immunoprophylaxis. In Roizman, B. and Lopez, C. (eds.). *The Herpesviruses*. Vol. 4, pp. 201-27. (New York and London: Plenum Press)
3. Betts, R. F. (1984). The relationship of epidemiology and treatment factors to infection and allograft survival in renal transplantation. In Plotkin, S. A., Michelson, S., Pagano, J., and Rapp, F. (eds.). *CMV: Pathogenesis and Prevention of Human Infection*. pp. 87-101. (New York: Alan R. Liss)
4. Nowak, B., Sullivan, C., Sarnow, P., Thomas, R., Bricout, F., Nicolas, I. C., Fleckenstein, B. and Levine, A. J. (1984). Characterization of monoclonal antibodies and polyclonal immune sera directed against human cytomegalovirus virion proteins. *Virology*, **132**, 807-13
5. Pereira, L., Hoffman, M., Tatsuno, M. and Dondero, D. (1984). Polymorphism of human cytomegalovirus glycoproteins characterized by monoclonal antibodies. *Virology*, **139**, 73-86
6. Britt, W. (1984). Neutralizing antibodies detect a disulfide-linked glycoprotein complex within the envelope of human cytomegalovirus. *Virology*, **135**, 369-78

7. Rasmussen, L., Mullenax, I., Nelson, R. and Merigan, T. C. (1985). Viral polypeptides detected by a complement-dependent neutralizing murine monoclonal antibody to human cytomegalovirus. *J. Virol.*, **55**, 274–80

8. Benko, D. M. and Gibson, W. (1986). Primate cytomegalovirus glycoproteins: Lectin-binding properties and sensitivities to glycosidases. *J. Virol.*, **59**, 703–13

9. Farrar, G. H. and Oram, J. D. (1984). Characterization of the human cytomegalovirus envelope glycoproteins. *J. Gen. Virol.*, **65**, 1991–2001

10. Britt, W. J. and Auger, D. (1986). Synthesis and processing of the envelope gp55-116 complex of human cytomegalovirus. *J. Virol.*, **58**, 185–91

11. Young, R. and Davis, R. W. (1983). Efficient isolation of genes by using antibody probes. *Proc. Natl. Acad. Sci. USA*, **80**, 1194–8

12. Mach, M., Utz, U. and Fleckenstein, B. (1986). Mapping of the major glycoprotein gene of human cytomegalovirus. *J. Gen. Virol.*, **67**, 1461–7

13. Heilbronn, R., Jahn, G., Bürkle, A., Freese, U.-K., Fleckenstein, B. and zur Hausen, J. (1987). Genomic localization, sequence analysis and transcription of the putative human cytomegalovirus DNA polymerase gene. *J. Virol.* (In press)

14. Baer, R., Bankier, A. T., Biggin, M. D., Deininger, P. L., Farrell, R. B., Gibson, T. I., Hatfull, G., Hudson, G. S., Satchwell, S. C., Sequin, C., Tuffnell, P. S. and Barrell, B. G. (1984). DNA sequence and expression of the B95-8 Epstein-Barr virus genome. *Nature*, **310**, 207–11

15. Thorley-Lawson, D. A. and Geilinger, K. (1980). Monoclonal antibodies against the major glycoprotein (gp350/220) of Epstein-Barr virus neutralize infectivity. *Proc. Natl. Acad. Sci. USA*, **77**, 5307–11

16. Davison, A. J. and Scott, J. E. (1986). The complete DNA sequence of varicella zoster virus. *J. Gen. Virol.*, **67**, 1759–816

17. Keller, P. M., Davison, A. J., Lowe, R. S., Bennett, C. D. and Ellis, R. W. (1986). Identification and structure of the gene encoding gpII, a major glycoprotein of varicella zoster virus. *Virology*, **152**, 181–91

18. Pellett, P. E., Kousoulas, K. G., Pereira, L. and Roizman, B. (1985). Anatomy of the herpes simplex virus 1 strain F glycoprotein B gene: Primary sequence and predicted protein structure of the wild type and of monoclonal antibody-resistant mutants. *J. Virol.*, **53**, 243–53

19. Quinn, J. P. and McGeoch, D. J. (1985). DNA sequence of the region in the genome of herpes simplex virus type 1 containing the genes for DNA polymerase and the major DNA binding protein. *Nucl. Acid. Res.*, **13**, 8143–63

20. Spear, P. (1985). Glycoproteins specified by herpes simplex virus. In Roizman, B. and Lopez, C. (eds.). *The Herpesviruses*. Vol. 3, pp. 315–55. (New York and London: Plenum Press)

21. Waner, J. L. and Weller, T. H. (1978). Analysis of antigenic diversity among human cytomegalovirus by kinetic neutralization tests with high titered rabbit antiserum. *Infect. Immun.*, **21**, 151–7

22. Kilpatrick, B. A., Huang, E.-S. and Pagano, J. S. (1976). Analysis of cytomegalovirus genomes with restriction endonuclease *Hind*DIII and *Eco*RI. *J. Virol.*, **18**, 1095–105

23. Talbot, P. and Almeida, J. D. (1977). Human cytomegalovirus: purification of enveloped virions and dense bodies. *J. Gen. Virol.*, **36**, 345–9

10
Rubella Virus and its Effects on the Developing Nervous System

JERRY S. WOLINSKY

INTRODUCTION

Rubella was initially recognized as a distinct disease entity in the early nineteenth century[1]. For nearly 150 years, rubella virus was considered to cause only a mild morbilliform rash, occasional fever and a rather predictable lymphadenopathy. Predominantly a childhood disease, rubella is known to be endemic throughout the world and is associated with epidemics which occur in irregular fashion.

An observation of fundamental importance was made in 1941 when Norman Gregg reported an apparent epidemic of congenital cataracts[2]. Gregg, an astute ophthalmologist, encountered an exceptional number of children with cataracts, many of whom had additional serious congenital defects. He noted that this flurry of cases was directly preceded by a large rubella outbreak. Thus he suggested that the cataracts and the often associated congenital cardiac abnormalities were the consequence of natural rubella virus infections. These observations were initially met with heated scepticism. Nevertheless, as other patient populations were studied, it became clear that rubella virus could have devastating effects upon the unborn fetus when acquired by women during early pregnancy[3]. Thus, Gregg not only recognized the ravaging potential of rubella, an otherwise innocuous disease, he also introduced the concept of viruses as teratogenic agents.

Rubella virus is a member of the togavirus family. This family of viruses consists of a group of small agents that contain a ribonucleic acid (RNA) genome and a lipid envelope or 'toga'[4]. The togaviruses are divided into three genera based on serology[5,6]. The alphavirus genus consists of 26 different arthropod-transmitted viruses which cause diseases in man ranging from encephalitis (for example Central European, Western and Eastern encephalitis viruses) to fever, rash and arthritis (as with Sindbis and Ross River viruses)[5]. The arboviruses characteristically cause a rapidly lytic infection in mammalian cells, but typically produce a less destructive and more persistent infection in the cells of their non-vertebrate hosts. Ross River, Semliki Forest and Venezuelan equine viruses are notable in that they can produce demyelinating disease in laboratory models[7-9]. The pestivirus genus includes three viruses, border disease virus of sheep, bovine viral diarrhoea virus (BVDV) and hog cholera virus (HCV). These non-arthropod-borne viruses are of particular interest because all three pestiviruses can be transmitted to the developing fetus, and two of the viruses (BVDV and HCV) can cause congenital malformations in their natural host in a fashion analogous to rubella virus[10-12]. Rubella virus is the only member of the genus rubivirus[13]. Unlike most other togaviruses, rubella virus has no known invertebrate host and the only known natural reservoir for rubella virus is man.

To best appreciate the effects of rubella virus on the developing nervous system and fetal tissue in general it seems appropriate first to review selected aspects of the biology and molecular composition of rubella virus and the common illnesses that arise from or complicate acute rubella infections, before focusing directly on the congenital rubella syndrome (CRS) and its more remote consequences such as progressive rubella panencephalitis (PRP).

RUBELLA VIRUS

Considerable progress has been made in the characterization of rubella virus since its first isolation in culture 25 years ago[14, 15]. Rubella virus is a spherical and rather small particle which measures 50-70 nm in diameter[16, 17]. Individual virions consist of a 30 nm electron-dense core surrounded by a lipid envelope[18, 19]. A distinctive electron-lucent zone is found between the virus core and envelope which serves to distinguish rubella virus from the other togaviruses[20]. The virus contains a message or positive sense, single-stranded ribonucleic acid (RNA) genome. Although physicochemically very closely related to the alphaviruses, rubella virus does not show serological cross-reaction with any other member of the togavirus family.

Replication of rubella virus in tissue culture is seldom attended by observable cytopathic effects and intracellular inclusions are rarely seen. Viral infection in culture proceeds with limited or no alteration of host cell protein synthesis[21-23]. Under experimental conditions acute infection of cells is blocked by interferon[24] and amantadine[25]. Persistent infections of cell cultures can be readily established in vitro[26] and the growth rate of chronically infected cells is slowed[27]. The persistent infection of cell cultures by rubella virus has been explained by mechanisms involving temperature-sensitive mutants[28] and by the presence of defective interfering particles[29, 30]. It is unknown whether these in vitro findings pertain to rubella virus infections in man.

Converging information from a number of independent laboratories confirms that rubella virus is composed of three distinct structural polypeptides[17, 31-36]. Two of these are acylated glycoproteins[37] that are embedded in the envelope of the virion and project from it to form poorly resolvable 6-8 nm surface spikes[38]. This envelope can be selectively released from the virus by detergents[17, 33], and the resulting E1- and E2-containing rosettes maintain the organization of the spikes[17]. The larger surface glycoprotein, termed E1, has a reported molecular weight of 58 000-62 000 and is the virus haemagglutinin. This molecule contains monoclonal antibody-defined domains which are independently involved in virus attachment to the surface of red blood cells and the initiation of infection[35,39-43]. Monoclonal antibodies to two distinct domains on the E1 glycoprotein of the Therien strain of rubella virus that neutralize infectivity are also able to neutralize a panel of diverse rubella virus clinical isolates as well as the RA27/3 vaccine strain (J. S. Wolinsky, M. N. Waxham and E. Bumoivici-Klein, unpublished observations).

The second glycoprotein, E2, is a family of related molecules that range from 42 000–47 000 to 47 000–54 000 in molecular weight[33,35,44]. These derive from a single apoprotein[45] and appear to differ only in their extent of glycosylation[37]. The function of the E2 glycoprotein remains uncertain. At least a portion of E2 must be exposed at the surface of the virus as E2 is accessible to labelling by sodium [^{3}H]borohydride after treatment with galactose oxidase[33]. Additionally, one domain on E2 may be associated with virus infectivity[46] and strain-specific antigens likely reside on this molecule[47]. Nearly all of the E2 present in intact virus exists as a disulphide-bond-stabilized heterodimer with E1[32,35]. E1 is also present within mature virus as an E1–E1 homodimer and in monomeric form[35]. Thus, the available structural and functional data on rubella virus suggest that E1 is the dominant surface molecule of the virus particle, and that E1 is probably the main target for the detection and subsequent elimination of rubella virus by the host's immune system.

The 30 nm central core of the virion contains a single copy of the 40S RNA genome[48,49] protected by multiple homodimers of the non-glycosylated 33 000–38 000 molecular weight capsid or C protein[33,35]. The genome of rubella virus contains approximately 11 000 bases of genetic information[44,48,50,51]. The naked genome is infectious but inefficiently so[51]. Once within cells, the genomic RNA can serve as a direct template for protein translation and thus the production of those non-structural proteins necessary for virus replication. As yet, little is known about the non-structural proteins of rubella virus. By analogy with the more well-studied alphaviruses, these non-structural proteins are likely to include molecules responsible for virus replication, transcription, and perhaps protease activities[52]. This process must give rise to the RNA polymerases required for the creation of negative strand templates needed for the generation of both subgenomic 24S RNA and full length progeny genomes. The subgenomic mRNA consists of approximately 3 500 bases, corresponds to the 3' one-third of the larger mRNA, and encodes the genetic information for the three structural proteins of the virus[44,49]. Rubella virus does not code for or carry a reverse transcriptase, and therefore has no ready mechanism for stably incorporating its genomic information into the gene pool of the host cell.

The synthesis and processing of the structural proteins of rubella virus proceeds through a high molecular weight polyprotein precursor. Translation of the 24S mRNA results in an 110 000 dalton polyprotein precursor (p110) whose gene order is NH$_2$-C-E2-E1-COOH[44,49,53]. The full length polyprotein precursor must initially be cleaved during or immediately following translation since p110 is not readily detected within the infected cell using short pulse metabolic labelling techniques. By analogy with Sindbis virus, an autoprotease activity of the C protein may be responsible for this early event[54]. The translation of E2 and E1 from subgenomic mRNA probably occurs on endoplasmic reticulum-bound ribosomes. After their synthesis, a number of modifications are made to the rubella virus glycoproteins before they are incorporated into the virus particle. Both E1 and E2 are altered by the addition of carbohydrates, predominantly mannose and glucosamine[31,33-35,37,55]. Additionally, both glycoproteins covalently incorporate [^{3}H]palmitic acid[37], a modification believed to be important in anchoring the glycoproteins within the lipid bilayer of the virion[56].

The recent cloning and sequencing of the rubella virus E1 gene confirms and extends the above information[57]. These show that the E1 protein consists of 481 amino acids. A putative signal sequence immediately precedes the amino terminus of E1[57], which provides a mechanism for the insertion of newly-synthesized molecules into the lumen of the endoplasmic reticulum[58]. Included within the E1 sequence are three N-linked glycosylation sites, and a possible transmembrane region near the carboxyl terminal end of the protein. Importantly, no homology is found when comparing the sequence of E1 with the analogous region of several alphaviruses[57].

Maturation of rubella virus occurs when the RNA-containing nucleocapsid core buds from a virus-modified cellular membrane, thus acquiring a viral envelope consisting of host cell lipids and the viral proteins E1 and E2. The budding process may commence from intracellular membranes (Golgi, mitochondria, endoplasmic reticulum) or from the plasma membrane of cells grown in culture[59]. The capacity of rubella virus to mature intracellularly and thus avoid detection by the host immune system may be an important determinant of its propensity to establish persistent infections in man.

The nature of the host cell receptor(s) for rubella virus is completely unknown. It is possible that the expression of this putative receptor is phylogenetically restricted. If true, this would provide a plausible explanation of the narrowly restricted host range of rubella virus. It is also conceivable that expression of the rubella virus receptor is restricted amongst the various specialized cell types of human tissues, and that the distribution of the receptor-bearing cells might explain some aspects of the pathogenesis of rubella-related diseases. Several approaches are presently being used in an attempt to identify and isolate the rubella virus receptor using either virus or anti-idiotypic antibody as suitable receptor ligands[60]. The relatively poor yields of rubella virus in culture have complicated the former approach, a limitation which may be overcome through the use of cDNA clones of the E1 gene region inserted into appropriate expression vectors. Encouraging initial results have been obtained with polyclonal xenogeneic and monoclonal syngeneic anti-idiotypic antibodies generated to mouse monoclonal antibodies which define two distinct domains on the E1 glycoprotein (J. S. Wolinsky, B. L. Slagle and A. Nath, unpublished observations). Antibodies to these two domains (E1$_c$ and E1$_d$) block virus infectivity without interfering with virus adsorption to red blood cells, implying that these regions of the E1 glycoprotein are directly involved in critical early interactions of the virus with the cell plasma membrane that lead to virus entry into the host cell. The antibodies generated thus far fulfil most of the criteria of anti-idiotypes of the 'internal image' type[61]. Further analysis should clarify whether or not these reagents will provide the necessary ligands to access the rubella virus receptor.

Rubella

The infection caused by rubella virus in early childhood is usually mild. It is characterized by any combination of symptoms which include maculo-papular rash, lymphadenopathy, low-grade fever, coryza, conjunctivitis, sore throat and arthralgia. Subclinical primary infections may occur in approximately 25% of susceptible individuals exposed to rubella[62, 63]. In industrialized countries most children are affected at between 5 and 7 years of age[64, 65]. Infection of adolescents and young adults is not uncommon, particularly in situations where a large number of susceptible individuals are housed closely together[66]. Seasonal peaks of infection occur in the winter months but the virus must circulate year-round at low levels in man, as man is the only known natural host. The disease occurs in infrequent epidemics every 5–9 years with major epidemics occurring about every 30 years[67]. The last major epidemic in the United States occurred in 1964 prior to the introduction of several attenuated live virus vaccines[68].

The mucosa of the upper respiratory tract and nasopharyngeal lymphoid tissue serve as portals of entry and the initial sites of virus replication. An incubation period of about 7 days is followed by lymphocyte-borne viraemia[69] and nasopharyngeal viral shedding. Shed virus serves as the source of contagion for susceptible individuals. The viraemia precedes the development of the rash by several days and ceases shortly after the rash appears[70]. The viraemic phase may be marked by mild prodromal symptoms of malaise. The rash is maculopapular with few specific features although it almost invariably begins on the face at 16 to 21 days post infection before becoming generalized. Posterior cervical adenopathy is characteristic. The entire clinical syndrome clears rapidly in days and is rarely attended by more significant symptoms.

A serological response is measurable at the onset of the rash and con-tinues to evolve over the next few weeks[62, 71–73]. The presence of these anti-bodies results in a rapid decline of the viraemia. Since the appearance of the rubella rash coincides with the detection of rubella-specific antibody, it has been postulated that the rash is an immune-complex-mediated phenomenon. Virus can be isolated from skin biopsies at the time of rash[74]. The initial rubella-specific antibody response is of the IgM type and these antibodies may persist for as long as a year after acute infection[75]. Low levels of haemagglutination inhibition and complement fixation antibody of the IgG class persist indefinitely after childhood infection. Depending on the assay used, immune complexes can be found with modest frequency in human sera, especially in association with a variety of viral infections. Immune complexes containing rubella-specific antibody or antigen are frequently found following rubella infections[76–78]. However, rubella-specific immune complexes are unusual late after natural or vaccine-induced rubella infections in the absence of persistent symptoms[77].

Transient depression of lymphocyte responsiveness to mitogenic stimulation follows natural or vaccine infections in children and adults[79,80]. However, specific immune responses do develop that can be measured *in vitro* and include proliferative responses, lymphocyte-mediated cytotoxicity, and lymphokine secretion[81-84]. Re-infection is possible but uncommon after naturally acquired rubella[85,86].

While the clinical manifestations of natural rubella are typically mild certain complications may arise. Central nervous system complications are distinctly uncommon, occurring in less than one in 5000 cases. Neural syndromes usually take the form of an acute toxic encephalopathy or encephalomyelitis[87-89]. Acute polyarthritis frequently develops following natural rubella infection[71,90] or vaccination[91,92] and may involve any joint, with the hands being the most commonly affected[93]. This problem may be particularly frequent amongst young women[92]. The usually transient arthralgia may persist or recur in some patients over years[91]. The precise mechanism by which rubella virus is involved in rubella arthritis is unknown but most certainly includes a persistent infection by the virus. Rubella virus can be isolated from synovial fluid for years following natural infection on immunization[90,93-95], and has been recovered from circulating mononuclear cells of children with chronic arthritis[96]. Additionally, human synovial cells cultured *in vitro* can be persistently infected with rubella virus[97].

The most devastating consequences of natural rubella are the abortions, miscarriages, stillbirths and fetal malformations that arise from maternal infection during the first trimester of pregnancy. Once rubella virus was isolated in tissue culture[14,15], successful vaccine development soon followed. Unfortunately, the introduction of the vaccine in 1969 was too late to prevent an epidemic of rubella in the United States. In this epidemic an estimated 20 000 infants suffered permanent damage from *in utero* exposure to rubella. Prevention of such devastation was the primary goal for development of a rubella vaccine.

The development of rubella vaccines[98,99] created a unique situation in which the primary group of people who contract and spread natural rubella (school-aged children) differed from the group who would most benefit from the vaccine (unborn children of pregnant women). Thus ethical considerations led to divergent vaccination policies in different countries. Consequently, there have been markedly varied initial results in achievement of the vaccine's key goal. The United States chose to disrupt the natural spread of the disease by mandatory vaccination of all pre-school children. Since vaccine introduction, the incidence of congenital rubella has declined remarkably in the United States with only two indigenous cases reported in 1985 to the National Congenital Rubella Register[100]. However, isolated outbreaks of rubella amongst young adults continue to be of concern[101]. In contrast, the United Kingdom undertook a policy of vaccinating 10-15-year-old girls only, with revaccination of seronegative women of childbearing age[102,103]. The number of congenital rubella cases has diminished but natural rubella infection is still prevalent among boys and prepubescent girls in the

United Kingdom[104,105]. Therefore, all women of childbearing age must continue to be screened for the presence of protective antibodies to rubella virus.

The vaccine strain currently licensed for distribution in the United States is designated RA27/3[98]. The RA27/3 isolate was attenuated by multiple passages on WI-38 human diploid cell culture, and thus avoids the possible side effects associated with vaccine viruses grown in non-human cells. The virus induces protective antibodies in 95% of vaccinated individuals. Vaccinees usually show no overt signs of disease, and transmission of vaccine virus to susceptible individuals is not observed[106]. However, the current vaccine virus is not without problems; 10–40% of vaccinated adult women develop an acute, usually transient, arthritis[107]. The vaccine virus can persist in these vaccines and has been recovered from one patient 2 years following immunization[93]. The long-term effects of such viral persistence are unknown. Although vaccine-induced antibody titres are generally lower than those following a natural infection, the level appears sufficient to protect against subsequent symptomatic infection. Re-infection in the presence of natural or immunization-induced low level antibody is rare. Usually such re-infections do not result in viraemia (with subsequent fetal infection)[108], although rare possible exceptions have been reported[109,110].

A very serious concern of the current rubella vaccination programme has been that of the risk to the developing fetus in mothers who are immunized during early pregnancy. In such instances, the placenta and fetal tissue may become infected[111,112]. However, none of the children born of mothers immunized during pregnancy have malformations compatible with the CRS[113]. Thus, the RA27/3 strain of rubella virus appears to lack the teratogenic capacity of natural (wild-type) rubella virus. Since a theoretical risk to the fetus still exists, pregnancy is considered a contraindication to rubella vaccination. However, inadvertent vaccination of a pregnant woman is not generally considered reason enough routinely to terminate a pregnancy.

CONGENITAL RUBELLA SYNDROME

In contrast to acquired infections, gestational rubella can have dire consequences for fetal development[2,114,115]. The placenta is frequently infected during maternal viraemia with subsequent dissemination of virus to the fetus[116,117]. Virus spreads widely in the developing fetal tissues and almost any organ may be infected[118]. Detailed analysis of rubella virus-infected products of conception shows consistent fine structural changes in the endothelial lining of blood vessels of brain and other organs, as well as a distinct lack of inflammatory changes[119]. While a proportion of infected fetuses may escape damage, pathology often ensues following infection before the sixteenth week of gestation, and 80% of neonates show sequelae of infections which begin in the first trimester[120]. Sometimes fetal rubella infection in early gestation causes no observable damage even though there is serological evidence of infection[121,122].

Unfortunately, precise mechanisms of fetal teratogenesis remain unknown. A favoured hypothesis is that the direct effects of virus replication on individual cells and their progeny during critical stages of the ontogeny of specific fetal organs gives rise to the wide range of abnormalities that together comprise the CRS[115,123]. It has been recently shown that human embryonic cells persistently infected *in vitro* with rubella virus display an altered responsiveness to the growth-promoting properties of epidermal growth factor, as well as a decreased capacity for collagen synthesis[124]. Thus, it is possible that a non-cytopathic rubella virus infection of embryonic cells *in vivo* may upset the normally delicate balance of cellular growth and differentiation. The report of successful non-lytic infection of mouse blastocysts by rubella virus[125] hopefully will lead to an animal model for studying mechanisms of viral teratogenesis.

Common manifestations of the CRS include glaucoma, cataract, hepatosplenomegaly, pulmonary hypertension, patent ductus arteriosus, thrombocytopenia purpura, bony radiolucencies and excretion of high amounts of virus from nasopharynx and urine for several weeks or months of postnatal life[126]. The overwhelming majority of CRS children (80%) show some type of CNS involvement[127,128]. Clinical manifestations expressed in infancy include bulging anterior fontanelle, lethargy, irritability and motor tone abnormalities. Residual signs in survivors include mental retardation, motor disabilities, abnormal posture and movements, and neurosensory hearing loss. Laboratory findings may include an increased cerebrospinal fluid (CSF) protein and abnormal electroencephalograms. Virus has been isolated from approximately 30% of CSF samples[127]. Neuropathological findings at autopsy of symptomatic newborns and infants differ from those found in early products of gestation. These include vascular abnormalities and focal areas of parenchymal and perivascular necrosis[128,129]. Damage of large vessels is sometimes accompanied by ischaemic necrosis of adjacent CNS tissue[129]. Inflammatory changes in the brain are variably observed, having been noted in one study[127], but lacking in others[129].

Typically, rubella virus can be isolated from almost any organ of full-term infants and may be recovered for one year or more in many surviving CRS infants[116,127,130–135]. Therefore, neither the transferred maternal nor developing fetal immune response is sufficient to eliminate the virus *in utero* even though maternal IgG and fetal IgM can neutralize the virus *in vitro*[114]. The mechanism by which rubella virus escapes immune elimination and establishes a persistent infection in these congenitally infected infants remains unknown. Those CRS infants infected during early gestation often demonstrate an impaired rubella-specific cell-mediated immune response[136]. Such infants produce detectable rubella-specific IgM and IgG[137]. But these titres can fall rapidly and antibody may disappear[138,139]. Late re-infection appears possible[140]. Furthermore, the humoral immune response of CRS infants differs qualitatively from that of naturally infected or immunized persons. Whereas the latter groups have antibodies against each of the three structural proteins of the virus, CRS infants often lack antibodies to the C protein and demonstrate weak reactivity to the E2 protein[141]. The reason for this abnormal humoral immune response in CRS infants is unclear. It may reflect an immature immune system of the infants, strain differences of rubella viruses in nature, or restricted virus expression.

Circulating immune complexes, composed in part of rubella-specific immunoglobulins, are found in sera from nearly half of all infants with the CRS[77]. Rubella antigen-containing immune complexes have been found in the sera of neonates with the late CRS[142], and there is an increased likelihood of finding rubella antibody containing immune complexes in the serum of congenital rubella children with active clinical problems[77]. These observations make it attractive to suggest that rubella-specific immune complexes may be a marker for continued antigenaemia.

PROGRESSIVE RUBELLA PANENCEPHALITIS

While many of the effects of fetal rubella are manifest at birth, some defects such as mental retardation and hearing loss may not become clinically apparent for several years[127]. However, congenital rubella has generally been considered a non-progressive syndrome at least after the second or third year of life[132,143]. Late consequences of congenital rubella are now being appreciated as these children are carefully followed longitudinally[144]. Diabetes mellitus occurs as overt or latent disease in as many as 20% of young adults with the CRS[145,146]. The pancreas can be a target for viral replication in fetal life[147]. Similarly, an unusual late-onset rubella encephalitis has been described[148-150] variably referred to as PRP[148], chronic progressive panencephalitis[149] and non-congenital rubella encephalitis[151]. Like other slow virus diseases of the CNS, PRP is characterized by a prolonged asymptomatic period, followed by the onset of symptoms of neural deterioration including behavioural changes, intellectual decline, ataxia, spasticity, and sometimes seizures[152]. Ingravescent neurological dysfunction follows which ultimately proceeds to death. Although PRP has most often been associated with CRS patients, it may also be a very rare, late complication of natural childhood rubella[151,153-155]. Progressive rubella panencephalitis, as currently defined, appears to be a distinctly uncommon disease. Most cases have been recognized following congenital rubella in non-epidemic years and it is possible that additional cases will be seen as a legacy of the 1964 rubella epidemic.

The brain appears to be the exclusive site of clinical and pathological involvement in this disorder. As the disorder progresses, dementia becomes severe and mutism, spastic quadriparesis and evidence of brain-stem involvement with ophthalmoplegia develop. Multifocal myoclonus has been observed but in only one case has this been a prominent feature of the disorder[155]. Progressive rubella panencephalitis is not geographically restricted, having been observed in five countries on four continents. To date it appears to be exclusively a disorder of males.

Several lines of evidence link rubella virus to the aetiology of PRP. Both sera and CSF from PRP patients contain antibodies which react with each of the structural proteins of rubella virus[156]. Rubella-specific immune complexes have been demonstrated in sera from several PRP cases[157,158]. Rubella virus antigens have been visualized *in situ* within cells from CSF of a PRP patient[155]. Rubella virus has been isolated from peripheral blood mononuclear cells of a PRP patient[159]. Finally, the virus has been isolated from a PRP brain by co-cultivation techniques[160]. These data support the concept that chronic rubella virus infection is present in PRP.

The CSF protein electrophoresis has disclosed greatly increased proportions of gamma globulin which when analysed by agarose gel electrophoresis show an oligoclonal pattern[161]. Immunoglobulin bands analogous to those in CSF can be selectively adsorbed from serum with antigens extracted from rubella virus infected cells[162]. Both sera and CSF contain antibodies which react strongly with all of the structural proteins of rubella virus in immuno-precipitation assays[156]; a pattern which is unusual for sera obtained in other rubella-related syndromes[141]. Intrathecal production of the CSF rubella virus-specific antibodies is supported by application of the formula derived by Tourtellotte[149], the demonstration of immunoglobulin-containing plasma cells in brain tissue[161], and analysis of immunoglobulins extracted from brain[156]. The presence of rubella antigen as determined by indirect immuno-fluorescence of cells obtained from the CSF has been reported[155]. This is a new finding of considerable practical and theoretical significance if confirmed in subsequent cases.

Circulating immune complexes have been demonstrated in the sera of several cases using a variety of techniques[157,158]. When dissected by immuno-chemical means, the complexes have been shown to be composed, at least in part, of rubella-specific antibody of the IgG class and a component of the rubella virus E1 molecule[158]. These findings would imply that some form of viral replication is an ongoing process in PRP. This, and other evidence reviewed below, suggests that immune complexes may participate in some manner in disease pathogenesis.

Biopsy material shows a perivascular infiltration of lymphocytes and plasma cells which is more prominent in the white matter than the cortex[148,149]. Microglial nodules are present and some neuronal loss and gliosis is apparent. Periodic acid–Schiff positive deposits may be seen in vessels of the subcortical white matter; a finding which may be diagnostically useful. No intranuclear nor intracytoplasmic inclusions are seen. The activity of the process varies pathologically and biopsy specimens may only reflect chronic inflammatory change, with neuronal loss and gliosis in regions of brain that clinically have been involved for extended periods of time. Isolation of virus from brain biopsy material co-cultivated with appropriate cell lines has been successful[160], but negative isolation is frequent. Autopsy material yields remarkably uniform findings[148,163,164]. Coronal sections of brain show lateral ventricular dilation with marked, irregular shrinkage and gelatinous change in the adjacent white matter. The cerebellum is remarkably atrophic with marked secondary enlargement of the fourth ventricle. Microscopic evidence of subacute to chronic inflammatory change is diffuse in the leptomeninges, and perivascular cuffs composed of mononuclear lympho-cytes and plasma cells are found throughout the brain parenchyma. There is extensive destruction of white matter with loss of myelin, relative preservation of axons, and severe gliosis. Perivascular accumulations of glycoprotein, acid mucopolysaccharide and iron-containing material are numerous in the vessels of the hemispheric white matter, basal ganglia and cerebellum. Immunocytochemical studies show that these perivascular deposits contain

IgG, but lack significant detectable amounts of IgM, IgA or complement[164]. A histologically similar vascular change can also be seen in the CRS, but is not as extensive and white matter destruction is limited in the congenital cases[127,129,165]. Neuronal loss in cortex can be extensive. Cerebellar atrophy is severe with loss of Purkinje and granule cells, fibrillary gliosis, loss of myelin and disruption and some loss of axons. Electron microscopy has shown numerous tubular aggregates within endothelial cells and otherwise typical nuclear bodies within reactive astrocytes. Viral particles have not been identified. Search for the presence of viral gene products in PRP tissue has generally been unrewarding. However, preliminary data suggests the isolated presence of E2 localized to parenchymal vessels of brain tissues from a recently autopsied case using monoclonal antibodies and indirect immuno-fluorescent techniques (J. S. Wolinsky, unpublished observations); a finding which requires further confirmation.

Critical information which could be used to construct a logical and integrated hypothesis for the pathogenesis of PRP remains unavailable. However, any such proposal must take into account a number of well supported observations enumerated above. While there is no known mechanism available to readily explain the persistence of rubella virus, sequestration of the rubella genome in some state for extended intervals appears inescapable. The continued or renewed generation of rubella-specific gene products, if not complete virus, is likely during symptomatic persistent infection in order to provide the viral components of the circulating immune complexes found in PRP. Major defects in the host's immune system are unlikely to account for the syndrome, although selected cellular second mediators may be down-regulated by humoral factors such as the immune complexes. Moreover, the brisk humoral immune response observed may contribute to the immunopathology of PRP. The organ-specific vascular lesions of PRP strongly suggest local deposition of immunoglobulin, perhaps in response to antigens produced or deposited in relationship to the abluminal surfaces of capillaries or the foot process of neighbouring astrocytes. Finally, a rather selective loss of myelin and oligodendroglia, and the near complete destruction of Purkinje cells may speak for some degree of selective tropism for rubella virus within the CNS or at least a more profound and lethal effect on the metabolism of these cells when invaded by rubella virus.

SUMMARY

Perhaps the most remarkable feature of rubella virus is its implication in a diverse array of clinical diseases including natural rubella, rubella arthritis, rubella encephalopathy, CRS, CRS-associated juvenile onset diabetes, and PRP. The role of rubella virus in the pathogenesis of each of these diseases remains incompletely understood. Rubella probably results from the combined immediate effects of virus replication in tissues and the directed host response to viral-infected cells. Rubella arthritis may reflect a host immune response to rubella-infected synovial membrane. Rubella para-infectious encephalopathy may not directly involve the virus, but could rather represent a host response to some toxic product of natural rubella.

In contrast, rubella virus involvement in PRP is probably both a direct consequence of persistent virus replication and complex host responses.

Our understanding of disease mechanisms in the rubella virus-related syndromes is hindered by the current lack of a suitable animal model system that fully mimics the infection seen in man. Still, several features of rubella virus pathogenesis appear ripe for future research using recently available molecular probes. These include determining the mechanisms of viral persistence in man, the details of the teratogenesis of wild-type rubella, and the pursuit of a subunit vaccine.

Acknowledgements

Recent work from this laboratory was supported in part by a grant from the National Multiple Sclerosis Society.

References

1. Cooper, L. Z. (1985). The history and medical consequences of rubella. *Rev. Infect. Dis.*, 7, S2–S10
2. Gregg, N. M. (1941). Congenital cataract following German measles in the mother. *Trans. Ophthalmol. Soc. Aust.*, 3, 35–46
3. Wesselhoeft, C. (1947). Rubella (German measles). *N. Engl. J. Med.*, 236, 943–50
4. Fenner, F. (1976). Classification and nomenclature of viruses. *Intervirology*, 7, 1–115
5. Shope, R. E. (1985). Epidemiology: Mechanisms of cause, distribution and transmission of viral disease. In Fields, B. N. (ed.) *Virology*. pp. 145–52. (New York: Raven Press)
6. Westaway, E. G., Brinton, M. A., Gaidamouich, S. Y., Horzinek, M. C., Igarashi, A., Kaariainen, L., Luou, D. K., Porterfield, J. S., Russell, P. K. and Trent, D. W. (1985). Flaviviridae. *Intervirology*, 24, 183–92
7. Chew-Lim, M., Scott, T. and Webb, H. E. (1978). An ultrastructural study of cerebellar lesions induced in mice by three inoculations of avirulent Semliki Forest virus. *Acta Neuropathol.*, 41, 55–9
8. Dal Canto, M. C. and Rabinowitz, S. G. (1981). Central nervous system demyelination in Venezuelan equine encephalomyelitis infection. An experimental model of virus-induced myelin injury. *J. Neurol. Sci.*, 49, 397–418
9. Seay, A. R. and Wolinsky, J. S. (1983). Ross River virus-induced demyelination: II. Ultrastructural studies. *Ann. Neurol.*, 14, 559–68
10. Snowdon, W. A., Parsonson, I. M. and Broun, M. L. (1975). The reaction of pregnant ewes to inoculation with mucosal disease virus of bovine origin. *J. Comp. Pathol.*, 85, 241–51
11. Brinton, M. A. (1980). Non-arbo togaviruses. In Schlesinger, R. W. (ed.) *The Togaviruses*. pp. 623–66. (New York: Academic Press)
12. Potts, B., Elder, G., Sever, J., Sawyer, M. and Osburn, B. (1986). Border disease of sheep: An animal model for virus-induced hypomyelination. *Neurology*, 36, S204
13. Matthews, R. E. F. (1982). Classification and nomenclature of viruses. Third report of the International Committee on the Taxonomy of Viruses. *Intervirology*, 17, 1–199
14. Parkman, P. D., Buescher, R. L. and Arnstein, M. S. (1962). Recovery of rubella virus from army recruits. *Proc. Soc. Exp. Biol. Med.*, 111, 225–30
15. Weller, T. H. and Neva, F. A. (1962). Propagation in tissue culture of cytopathic agents from patients with rubella-like illness. *Proc. Soc. Exp. Biol. Med.*, 111, 215–25
16. Bardeletti, G., Kessler, N. and Aymard-Henry, N. (1975). Morphology. Biochemical analysis and neuraminidase activity of rubella virus. *Arch. Virol.*, 49, 175–86
17. Payment, P., Ajdukovic, D. and Pavilanis, V. (1975). Le virus de la ruberole. I. Morphologie et proteines structurales. *Can. J. Microbiol.*, 21, 703–9
18. Murphy, F. A., Halonen, P. E. and Harrison, A. K. (1968). Electron microscopy of the development of rubella virus in BHK-21 cells. *J. Virol.*, 2, 1223–7

19. Oshiro, L. S., Schmidt, N. J. and Lennette, E. H. (1969). Electron microscopic studies of rubella virus. *J. Gen. Virol.*, **5**, 205-10
20. Murphy, F. A. (1980). Togavirus morphology and morphogenesis. In Schlesinger, R. W. (ed.) *The Togaviruses.* pp. 241-316. (New York: Academic Press)
21. Maes, R., Vaheri, A., Sedwick, D. and Plotkin, S. (1966). Synthesis of virus and macromolecules by rubella virus infected cells. *Nature*, **210**, 384-6
22. Olson, G. B., Dent, P. B., Rawls, W. E., South, M. A., Montgomery, J. R., Melnick, J. L. and Good, R. A. (1968). Abnormalities of *in vitro* lymphocyte responses during rubella virus infections. *J. Exp. Med.*, **128**, 47-68
23. Chantler, J. K. (1979). Rubella virus: Intracellular polypeptide synthesis. *Virology*, **98**, 275-8
24. Potter, J. E., Banatavala, J. E. and Best, J. (1973). Interferon studies with Japanese and U.S. rubella strains. *Br. J. Med.*, **1**, 197-9
25. Maassab, H. F. and Cochran, K. W. (1964). Rubella virus: Inhibition *in vitro* with amantadine hydrochloride. *Science*, **145**, 1443-4
26. Rawls, W. E., Desmyter, J. and Melnick, J. L. (1968). Serologic diagnosis and fetal involvement in maternal rubella: Criteria for abortion. *J. Am. Med. Assoc.*, **203**, 627-31
27. Rawls, W. E. and Melnick, J. L. (1966). Rubella virus carrier cultures derived from congenitally infected infants. *J. Exp. Med.*, **123**, 795-816
28. Williams, M. P., Brawner, T. A. and Riggs, H. G. (1981). Characteristics of a persistent rubella infection in a human cell line. *J. Gen Virol.*, **52**, 321-8
29. Norval, M. (1979). Mechanism of persistence of rubella virus in LLC-MK2 cells. *J. Gen. Virol.*, **43**, 289-98
30. Bohn, E. M. and Van Alstyne, D. (1981). The generation of defective interfering rubella virus particles. *Virology*, **111**, 549-54
31. Vaheri, A. and Hovi, T. (1972). Structural proteins and subunits of rubella virus. *J. Virol.*, **9**, 10-16
32. Ho-Terry, L. and Cohen, A. (1980). Degradation of rubella virus envelope components. *Arch. Virol.*, **65**, 1-13
33. Oker-Blom, C., Kalkkinen, N., Kaariainen, L. and Pettersson, R. F. (1983). Rubella virus contains one capsid protein and three envelope glycoproteins, E1, E2a and E2b. *J. Virol.*, **46**, 964-73
34. Toivonen, V., Vainionpaa, R., Salmi, A. and Hyypia, T. (1983). Glycoproteins of rubella virus: Brief report. *Arch. Virol.*, **77**, 91-5
35. Waxham, M. N. and Wolinsky, J. S. (1983). Immunochemical identification of rubella virus hemagglutinin. *Virology*, **126**, 194-203
36. Bowden, D. S. and Westaway, E. G. (1984). Rubella virus: Structural and non-structural proteins. *J. Gen. Virol.*, **65**, 933-43
37. Waxham, M. N. and Wolinsky, J. S. (1985). A model of the structural organization of rubella virions. *Rev. Infect. Dis.*, **7**, S133-S139
38. Trudel, M., Ravaoarinoro, M. and Payment, P. (1980). Reconstitution of rubella hemagglutinin on liposomes. *Can. J. Microbiol.*, **26**, 899-904
39. Ho-Terry, L., Cohen, A. and Tedder, R. S. (1984). Immunological characterization of rubella virion polypeptides. *J. Med. Microbiol.*, **17**, 105-9
40. Waxham, M. N. and Wolinsky, J. S. (1985). Detailed immunologic analysis of the structural polypeptides of rubella virus using monoclonal antibodies. *Virology*, **143**, 153-65
41. Dent, P. B. and Rawls, W. E. (1971). Human congenital rubella: The relationship of immunologic aberation to viral persistence. *Ann. N.Y. Acad. Sci.*, **181**, 209-22
42. Umino, Y., Sato, T. A., Katow, S., Matsuno, T. and Sugiura, A. (1985). Monoclonal antibodies directed to E1 glycoprotein of rubella virus. *Arch. Virol.*, **83**, 33-42
43. Ho-Terry, L., Terry, G. M., Cohen, A. and Londesborough, P. (1986). Immunological characterization of the rubella E1 glycoprotein. *Arch. Virol.*, **90**, 145-52
44. Pettersson, R. F., Oker-Blom, C., Kalkkinen, N., Kallio, A., Ulmanen, J., Kaariainen, L. and Partanen, P. (1985). Molecular and antigenic characteristics and synthesis of rubella virus structural proteins. *Rev. Infect. Dis.*, **7**, S140-S149
45. Kalkkinen, N., Oker-Blom, C. and Pettersson, R. F. (1984). Three gene codes for rubella virus structural proteins E1, E2a and E2b. *J. Gen. Virol.*, **65**, 1549-57

46. Green, K. Y. and Dorsett, P. H. (1986). Rubella virus antigens: Localization of epitopes involved in hemagglutination and neutralization by using monoclonal antibodies. *J. Virol.*, **57**, 893–8
47. Dorsett, P. H., Miller, D. C., Green, K. Y. and Byrd, F. I. (1985). Structure and function of the rubella virus proteins. *Rev. Infect. Dis.*, **7**, S150–S156
48. Sedwick, W. D. and Sokol, F. (1970). Nucleic acid of rubella virus and its replication in hamster kidney cells. *J. Virol.*, **5**, 478–89
49. Oker-Blom, C., Ulmanen, I., Kaarianen, L. and Pettersson, R. F. (1984). Rubella virus 40S genome RNA specifies a 24S subgenomic mRNA that codes for a precursor to structural proteins. *J. Virol.*, **49**, 403–8
50. Wong, K. T., Robinson, W. S. and Merigan, T. C. (1969). Synthesis of viral-specific ribonucleic acid in rubella virus-infected cells. *J. Virol.*, **4**, 901–3
51. Hovi, T. and Vaheri, A. (1970). Infectivity and some physicochemical characteristics of rubella virus ribonucleic acid. *Virology*, **42**, 1–8
52. Schlesinger, M. J. (1985). Replication of togaviruses. In Fields, B. N. (ed.) *Virology*. pp. 1021–32. (New York: Raven Press)
53. Oker-Blom, C. (1984). The gene order for rubella virus structural proteins is NH_2-C-E2-E1-COOH. *J. Virol.*, **51**, 354–8
54. Alperti, G. and Schlesinger, M. J. (1978). Evidence for an autoprotease of Sinbis virus capsid protein. *Virology*, **90**, 336–69
55. Bowden, D. S. and Westaway, E. G. (1985). Changes in glycosylation of rubella virus envelope proteins during maturation. *J. Gen. Virol.*, **66**, 201–6
56. Schmidt, M. F. G. and Schlesinger, M. J. (1980). Relation of fatty acid attachment to the translation and maturation of vesicular stomatitis and Sindbis virus membrane glycoproteins. *J. Biol. Chem.*, **255**, 3334–9
57. Frey, T. K., Marr, L. D., Hemphill, M. L. and Dominguez, G. (1986). Molecular cloning and sequencing of the region of the rubella virus genome coding for glycoprotein E1. *Virology*, **154**, 228–32
58. Blobel, G. and Dobberstein, B. (1975). Transfer of proteins across membranes; I. Presence of proteolytically processed and unprocessed nascent immunoglobulin light chains on membrane-bound ribosomes of murine myeloma. *J. Cell. Biol.*, **67**, 835–51
59. Bardeletti, G., Tektoff, J. and Gauthern, O. (1979). Rubella virus maturation and production in two host cell systems. *Intervirology*, **11**, 97–103
60. Gaulton, G. N., Co, M. S., Royer, H. D. and Greene, M. I. (1985). Anti-idiotype antibodies as probes of surface receptors. *Mol. Cell. Biochem.*, **65**, 5–21
61. Nisonoff, A. and Lamoyi, E. (1981). Implications of the presence of an internal image of the antigen in anti-idiotypic antibodies: Possible application to vaccine production. *Clin. Immunol. Immunopathol.*, **21**, 397–406
62. Green, R. H., Balsamo, M. R., Giles, J. P., Krugman, S. and Mirick, G. S. (1965). Studies of the natural history and prevention of rubella. *Am. J. Dis. Child.*, **110**, 348–65
63. Sever, J. L., Hardy, J. B., Nelson, K. B. and Gilkeson, M. R. (1969). Rubella in the collaborative perinatal research study. II. Clinical and laboratory findings in children through 3 years of age. *Am. J. Dis. Child.*, **118**, 123–32
64. Krugman, S. (1965). Rubella: Clinical and epidemiological aspects. *Arch. Ges. Virusforsch.*, **16**, 477–82
65. Horstmann, D. M. (1976). Rubella. In Evans, A. S. (ed.) *Viral Infections of Humans: Epidemiology and Control*. pp. 409–27. (New York: Plenum Publishing)
66. Vesikari, T. (1972). Antibody response in rubella reinfection. *Scand. J. Infect. Dis.*, **4**, 11–16
67. Witte, J. J., Karchmer, A. W., Case, G., Herrmann, K. L., Abrutyu, E., Kassanoff, I. and Neill, J. S. (1969). Epidemiology of rubella. *Am. J. Dis. Child.*, **118**, 107–11
68. Bart, S. W., Stetler, H. C., Preblud, S. R., Williams, N. M., Orenstein, W. A., Bart, K. J., Hinman, A. R. and Herrmann, K. L. (1985). Fetal risk associated with rubella vaccine: An update. *Rev. Infect. Dis.*, **7**, S95–S102
69. Chandler, J. K. and Tingle, A. J. (1982). Isolation of rubella virus from human lymphocytes after acute natural infection. *J. Infect. Dis.*, **145**, 673–7
70. Schiff, G. M. and Sever, J. L. (1966). Rubella: Recent laboratory and clinical advances. *Prog. Med. Virol.*, **8**, 30–61

71. Heggie, A. D. and Robbins, F. C. (1969). Natural rubella acquired after birth: Clinical features and complications. *Am. J. Dis. Child.*, **118**, 12–17
72. Horstman, D. M., Pajot, T. G. and Liebhaber, H. (1969). Epidemiology of rubella: Subclinical infection and occurrence of reinfection. *Am. J. Dis. Child.*, **118**, 133–6
73. Herrmann, K. L. (1985). Available rubella serologic tests. *Rev. Infect. Dis.*, **7**, S108–S112
74. Heggie, A. D. (1978). Pathogenesis of the rubella exanthema: Distribution of rubella virus in the skin during rubella with and without rash. *J. Infect. Dis.*, **137**, 74–7
75. Al-Nakib, W., Best, J. M. and Banatvala, J. E. (1975). Rubella-specific serum and nasopharyngeal responses following naturally acquired and vaccine-induced infection. *Lancet*, **1**, 182–5
76. Pernice, W., Schmitz, H., Schindera, F., Behrens, F. and Sedlacek, H. H. (1979). Antigen-specific detection of immune complexes in patients with hepatitis B, influenza A and rubella. *Behring Inst. Mitt.*, **64**, 102–8
77. Coyle, P. K., Wolinsky, J. S., Buimovici-Klein, E., Moucha, R. and Cooper, L. Z. (1982). Rubella-specific immune complexes after congenital infection and vaccination. *Infect. Immun.*, **36**, 498–503
78. Ziola, B., Lund, G., Meurman, O. and Salmi, A. (1983). Circulating immune complexes in patients with acute measles and rubella virus infections. *Infect. Immun.*, **41**, 578–83
79. McMorrow, L. E., Vesikari, T., Walman, S. R., Giles, J. P. and Cooper, L. Z. (1974). Suppression of the response of lymphocytes to phytohemagglutinin in rubella. *J. Infect. Dis.*, **130**, 464–9
80. Buimovici-Klein, E., Vesikari, T., Santangelo, C. F. and Copper, L. Z. (1976). Study of the lymphocyte *in vitro* response to rubella antigen and phytohemagglutin by a whole blood method. *Arch. Virol.*, **52**, 323–31
81. Honeyman, M. C., Forrest, J. M. and Dorman, D. C. (1972). Cell-mediated immune response following natural rubella vaccination. *Clin. Exp. Immunol.*, **17**, 665–71
82. Steele, R. W., Hensen, S. A., Vincent, M. M., Fuccillo, D. A. and Bellanti, J. A. (1974). Development of specific cellular and humoral immune responses in children immunized with live rubella. *J. Infect. Dis.*, **130**, 449–53
83. Morag, A., Morag, B., Bernstein, J. M., Buetner, K. and Ogra, P. L. (1975). *In vitro* correlates of cell-mediated immunity in human tonsils after natural or induced rubella virus infections. *J. Infect. Dis.*, **131**, 409–16
84. Buimovici-Klein, E. and Cooper, L. Z. (1985). Cell-mediated immune response in rubella infections. *Rev. Infect. Dis.*, **7**, S123–S128
85. Horstmann, D. M., Pajot, T. G. and Liebhaber, H. (1969). Epidemiology of rubella: Subclinical infection and occurrence of re-infection. *Am. J. Dis. Child.*, **118**, 133–6
86. Horstmann, D. M., Liebhaber, H., LeBouvier, G. L., Rosenberg, D. A. and Halstead, S. D. (1970). Rubella: Reinfection of vaccinated and naturally immune persons exposed in an epidemic. *N. Engl. J. Med.*, **283**, 771–8
87. Miller, H. G., Stanton, J. B. and Gibbons, J. L. (1956). Para-infectious encephalomyelitis and related syndromes. *Q. J. Med.*, **25**, 427–505
88. Johnson, K. P., Wolinsky, J. S. and Ginsberg, A. H. (1978). Immune mediated syndromes of the nervous system related to viral infections. In Vinken, P. J. and Bruyn, G. W. (eds.) *Handbook of Clinical Neurology*, Vol. 34, pp. 391–434. (Amsterdam: North Holland)
89. Waxham, M. N. and Wolinsky, J. S. (1984). Rubella virus and its effects on the central nervous system. *Neurol. Clin.*, **2**, 367–85
90. Grahame, R., Armstrong, R. and Simon, N. (1983). Chronic arthritis associated with the presence of intrasynovial rubella virus. *Ann. Rheum. Dis.*, **42**, 2–13
91. Chantler, J. K., Ford, D. K. and Tingle, A. J. (1981). Rubella associated arthritis: Rescue of rubella virus from peripheral blood lymphocytes two years post vaccination. *Infect. Immun.*, **32**, 1274–80
92. Tingle, A. J., Allen, M., Petty, R. E., Kettyls, G. D. and Chantler, J. K. (1986). Rubella-associated arthritis. I. Comparative study of joint manifestations associated with natural rubella infection and RA27/3 rubella immunization. *Ann. Rheum. Dis.*, **45**, 110–14
93. Chantler, J. K., Ford, D. K. and Tingle, A. J. (1982). Persistent rubella infection and rubella-associated arthritis. *Lancet*, **1**, 1323–5
94. Grahame, R., Simons, N. A. and Wilton, J. M. A. (1981). Isolation of rubella virus from synovial fluid in five cases of seronegative arthritis. *Lancet*, **2**, 649–52

95. Ford, D. K., DeRoza, D. M. and Reid, G. D. (1982). Synovial mononuclear cell responses to rubella antigen in rheumatoid arthritis and unexplained persistent knee arthritis. *J. Rheum.*, **9**, 420-3

96. Chantler, J. K., Tingle, A. J. and Petty, R. E. (1985). Persistent rubella virus infection associated with chronic arthritis in children. *N. Engl. J. Med.*, **313**, 1117-23

97. Cunningham, A. L. and Fraser, J. R. E. (1985). Persistent rubella virus infection of human synovial cells cultured *in vitro*. *J. Infect. Dis.*, **151**, 638-45

98. Plotkin, S. A. (1969). Attenuation of RA27/3 rubella virus in WI-38 human diploid cells. *Am. J. Dis. Child.*, **118**, 178-85

99. Prinzie, A., Huggelen, C. and Gold, J. (1969). Experimental live attenuated rubella virus vaccine. *Am J. Dis. Child.*, **118**, 172-7

100. Center for Disease Control (1986). Rubella and congenital rubella syndrome—United States, 1984-1985. *Morbid. Mortal. Weekly Rep.*, **35**, 129-35

101. Miller, K. A. and Zager, T. D. (1984). Rubella susceptability in an adolescent female population. *Mayo Clin. Proc.*, **59**, 31-4

102. Hinman, A. R., Orenstein, W. A., Bart, K. J. and Preblud, S. R. (1983). Rational strategy for rubella vaccination. *Lancet*, **1**, 39-43

103. Tobin, J. O., Sheppard, S., Smithells, R. W., Milton, A., Noah, N. and Reid, D. (1985). Rubella in the United Kingdom, 1970-1983. *Rev. Infect. Dis.*, **7**, S47-S52

104. Clarke, M., Boustred, J., Schild, G. C., Seagroatt, V., Pollock, T. M., Finlay, S. E. and Barbara, J. A. J. (1979). Effect of rubella vaccination programme on serological status of young adults in United Kingdom. *Lancet*, **1**, 1224-6

105. Anderson, R. M. and Grenfell, B. T. (1985). Control of congenital rubella syndrome by mass vaccination. *Lancet*, **1**, 827-8

106. Halstead, S. B. and Diwan, A. J. (1971). Failure to transmit rubella virus vaccine. A close-contact study in adults. *J. Am. Med. Assoc.*, **215**, 634-6

107. Polk, B. F., Modlin, J. F. and White, J. A. (1982). A controlled comparison of joint reactions among women receiving one of two rubella vaccines. *Am. J. Epidemiol.*, **115**, 19-25

108. O'Shea, S., Best, J. M. and Banatvala, J. E. (1983). Viremia, virus excretion, and antibody response after challenge in volunteers with low levels of antibody to rubella virus. *J. Infect. Dis.*, **148**, 639-47

109. Partridge, J. W., Flewett, T. H. and Whitehead, J. E. M. (1981). Congenital rubella affecting an infant whose mother had rubella antibodies before conception. *Br. Med. J.*, **282**, 187-8

110. Eilard, T. and Strannegard, O. (1974). Rubella re-infection in pregnancy followed by transmission to the fetus. *J. Infect. Dis.*, **129**, 594-6

111. Vaheri, A., Vesikari, T., Oker-Blom, N., Parkman, P. D., Veronelli, J. and Robbins, T. C. (1972). Isolation of attenuated rubella-vaccine virus from human products of conception and uterine cervix. *N. Engl. J. Med.*, **286**, 1071-4

112. Hayden, G. C., Herrmann, K. L. and Buimovici-Klein, E. (1980). Subclinical congenital rubella infection associated with maternal rubella vaccination in early pregnancy. *J. Pediatr.*, **96**, 869-72

113. Preblud, S. R. and Williams, N. M. (1985). Fetal risk associated with rubella vaccine: Implications for vaccination of susceptible women. *Obstet. Gyn.*, **66**, 121-3

114. Cooper, L. Z., Ziring, P. R., Ockerse, A. B., Fedun, B. A., Kiely, B. and Krugman, S. (1969). Rubella: Clinical manifestations and management. *Am. J. Dis. Child.*, **118**, 18-29

115. Alford, C. A. (1976). Rubella. In Remington, J. S. and Klein, M. O. (eds.) *Infectious Diseases of the Fetus and Newborn*. pp. 71-106. (Philadelphia: W. B. Saunders)

116. Alford, C. A., Neva, F. A. and Weller, T. H. (1964). Virologic and serologic studies on human products of conception after maternal rubella. *N. Engl. J. Med.*, **271**, 1275-81

117. Terry, G. M., Ho-Terry, L., Warren, R. C., Rodeck, C. H., Cohen, A. and Rees, K. R. (1986). First trimester prenatal diagnosis of congenital rubella: A laboratory investigation. *Br. Med. J.*, **292**, 930-3

118. Bellanti, J. A., Artenstein, M. S., Olsen, L. C., Buesher, E. L., Luhrs, C. E. and Milstead, K. L. (1965). Congenital rubella. Clinicopathologic, virologic and immunologic studies. *Am. J. Dis. Child.*, **110**, 464-72

119. Tondury, G. and Smith, D. W. (1966). Fetal rubella pathology. *J. Pediatr.*, **58**, 867-79

140

120. Miller, E., Cradock-Watson, J. E. and Pollock, T. M. (1982). Consequences of confirmed maternal rubella at successive stages of pregnancy. *Lancet*, **2**, 781-4
121. Dudgeon, J. A., Butler, N. R. and Plotkin, S. A. (1964). Further serologic studies on the rubella syndrome. *Br. Med. J.*, **2**, 155-60
122. Butler, N. R., Dudgeon, J. A., Hayes, H., Peckham, G. S. and Wybar, K. (1965). Persistence of rubella antibody with and without embryopathy. A follow-up study of children exposed to maternal rubell. *Br. Med. J.*, **2**, 1027-9
123. Rawls, W. E. (1974). Viral persistence in congenital rubella. *Prog. Med. Virol.*, **18**, 273-88
124. Yoneda, T., Urade, M., Sakuda, M. and Miyazaki, T. (1986). Altered growth, differentiation, and responsiveness to epidermal growth factor of human embryonic mesenchymal cells of palate by persistent rubella virus infections. *J. Clin. Invest.*, **77**, 1613-21
125. Hearn, A. M., O'Sullivan, M. A. and Atkins, G. J. (1986). Infection of cultured early mouse embryos with Semliki Forest and rubella viruses. *J. Gen. Virol.*, **67**, 1091-8
126. Dudgeon, J. A. (1969). Congenital rubella: Pathogenesis and immunology. *Am. J. Dis. Child.*, **118**, 35-44
127. Desmond, M. M., Wilson, G. S., Melnick, J. L., Singer, D. B., Zion, T. E., Rudolph, A. J., Pineda, R. G., Ziai, M. and Blattner, R. J. (1967). Congenital rubella encephalitis: Course and early sequelae. *J. Pediatr.*, **71**, 311-31
128. Desmond, M. M., Montgomery, J. R. and Melnick, J. L. (1969). Congenital rubella encephalitis. *Am. J. Dis. Child.*, **118**, 30-31
129. Rorke, L. B. (1973). Nervous system lesions in the congenital rubella syndrome. *Arch. Otolaryngol.*, **98**, 249-51
130. Monif, G. R. C. and Sever, J. L. (1966). Chronic infection of the central nervous system with rubella virus. *Neurology*, **16**, 111-12
131. Singer, D. B., Rudolph, A. J. and Rosenberg, H. S. (1967). Pathology of the congenital rubella syndrome. *J. Pediatr.*, **71**, 665-75
132. Menser, M. A., Dods, L. and Harley, J. D. (1967). A twenty-five year follow-up of congenital rubella. *Lancet*, **2**, 1347-50
133. Simons, M. J. and Jack, I. (1968). Lymphocyte viraemia in congenital rubella. *Lancet*, **2**, 953-4
134. Driscoll, S. G. (1969). Histopathology of gestational rubella. *Am. J. Dis. Child.*, **118**, 44-53
135. Hardy, J. (1973). Clinical and development aspects of congenital rubella. *Arch. Otolaryngol.*, **98**, 230-6
136. Buimovici-Klein, E., Lang, P. B., Ziring, P. R. and Cooper, L. Z. (1979). Impaired cell-mediated immune response in patients with congenital rubella. Correlation with gestational age at time of infection. *Pediatrics*, **64**, 620-6
137. Enders, G. (1985). Serologic test combinations for safe detection of rubella infections. *Rev. Infect. Dis..*, **7**, S113-S122
138. Cooper, L. Z., Florman, A. L., Ziring, P. R. and Krugman, S. (1971). Loss of rubella hemagglutination inhibition antibody in congenital rubella. Failure of seronegative children with congenital rubella to respond to HPV-77 vaccine. *Am. J. Dis. Child.*, **122**, 397-403
139. Ueda, K., Nishida, Y., Oshima, K., Yoshikawa, H., Ohashi, K. and Nomaka, S. (1975). Seven year follow-up study of the rubella syndrome in Ryukyu with special references to the persistence of rubella hemagglutination inhibition antibodies. *Jpn. J. Microbiol.*, **19**, 181-5
140. Menser, M. A., Slinn, R. F., Dods, L., Hertzberg, R. and Harley, J. D. (1968). Congenital rubella in a mother and son. *Aust. Paediatr. J.*, **4**, 200-2
141. DeMazancourt, A., Waxham, M. N., Nicholas, J. C. and Wolinsky, J. S. (1986). Antibody response to the rubella virus structural proteins in infants with the congenital rubella syndrome. *J. Med. Virol.*, **19**, 111-22
142. Tardieu, M., Grospierre, B., Durandy, A. and Griscelli, C. (1980). Circulating immune complexes containing rubella antigens in late-onset rubella syndrome. *J. Pediatr.*, **97**, 370-3
143. Peckham, C. S. (1972). Clinical and laboratory study of children exposed *in utero* to maternal rubella. *Arch. Dis. Child.*, **47**, 571-7
144. Sever, J. L., South, M. A. and Shaver, K. A. (1985). Delayed manifestations of congenital rubella. *Rev. Infect. Dis.*, **7**, S164-S169

145. Forrest, J. J., Menser, M. A. and Burgess, J. A. (1971). High frequency of diabetes mellitus in young adults with congenital rubella. *Lancet*, **2**, 332–4
146. Ginsberg-Fellner, F., Witt, M. E., Fedun, B., Doberson, M. J., McEvoy, R. C., Copper, L. Z., Notkins, A. L. and Rubinstein, P. (1985). Diabetes mellitus and autoimmunity in patients with the congenital rubella syndrome. *Rev. Infect. Dis.*, **7**, S170–S178
147. Cooper, L. Z., Green, R. H., Krugman, S., Giles, J. P. and Mirick, G. S. (1965). Neonatal thrombocytopenic purpura and other manifestations of rubella contracted *in utero*. *Am. J. Dis. Child.*, **110**, 416–27
148. Townsend, J. J., Baringer, J. R., Wolinsky, J. S., Malamud, N., Mednick, J. P., Panitch, H. S., Scott, R. A. T., Oshiro, L. S. and Cremer, N. E. (1975). Progressive rubella panencephalitis: Late onset after congenital rubella. *N. Engl. J. Med.*, **292**, 990–3
149. Weil, M. L., Itabashi, H. H., Cremer, N. E., Oshiro, L. S., Lennette, E. H. and Carnay, L. (1975). Chronic progressive panencephalitis due to rubella virus simulating subacute sclerosing panencephalitis. *N. Engl. J. Med.*, **292**, 994–8
150. Jan, J. E., Tingle, A. J., Donald, G., Kettyls, M., Buckler, W. S. J. and Dolman, C. L. (1979). Progressive rubella panencephalitis: Clinical course and response to 'Isoprinosine'. *Dev. Med. Child. Neurol.*, **21**, 648–52
151. Lebon, P. and Lyon, G. (1974). Non-congenital rubella encephalitis. *Lancet*, **2**, 468
152. Wolinsky, J. S. (1988). Progressive rubella panencephalitis. In Vinken, P. J. and Bruyn. G. W. (eds.) *Handbook of Clinical Neurology*. (In press) (Amsterdam: North-Holland)
153. Wolinsky, J. S., Berg, B. O. and Maitland, C. J. (1976). Progressive rubella panencephalitis. *Arch. Neurol.*, **33**, 722–3
154. Dayras, J. C., Lyon, G., Ponsot, G. and Allemon, M. C. (1980). L'encephalite chronique progressive de la rubeole: A propos d'une observation. *Semin. Hop. Paris*, **56**, 1703–8
155. Abe, T., Nakada, T., Hatanaka, H., Tajima, M., Hiraiwa, M. and Ushijima, H. (1983). Myoclonus in a case of suspected progressive rubella panencephalitis. *Arch. Neurol.*, **40**, 98–100
156. Wolinsky, J. S., Waxham, M. N., Hess, J. L., Townsend, J. J. and Baringer, J. R. (1982) Immunochemical features of a case of progressive rubella panencephalitis. *Clin. Exp. Immunol.*, **48**, 359–66
157. Weil, M. L., Perrin, L., Buimovici-Klein, E., Oldstone, M. and Cooper, L. Z. (1976). Immunologic abnormalities associated with chronic progressive panencephalitis due to congenital infection with rubella. *Clin. Res.*, **24**, 185
158. Coyle, P. K. and Wolinsky, J. S. (1981). Characterization of immune complexes in progressive rubella panencephalitis. *Ann. Neurol.*, **9**, 557–62
159. Wolinsky, J. S., Dau, P. C., Buimovici-Klein, E., Mednick, J., Berg, B. O., Lang, P. B. and Cooper, L. Z. (1979). Progressive rubella panencephalitis: Immunovirological studies and results of isoprinosine therapy. *Clin. Exp. Immunol.*, **35**, 397–404
160. Cremer, N. E., Oshiro, L. S., Weil, M. L., Lennette, E. H., Itabashi, H. and Carnay, L. (1975). Isolation of rubella virus from brain in chronic progressive panencephalitis. *J. Gen. Virol.*, **29**, 143–53
161. Wolinsky, J. S. (1978). Progressive rubella panencephalitis. In Vinken, P. J. and Bruyn, G. W. (eds.) *Handbook of Clinical Neurology*. Vol. 34, pp. 331–41. (Amsterdam: North-Holland)
162. Vandvik, B., Norrby, E., Steen-Johnson, J. and Stensvold, K. (1978). Progressive rubella panencephalitis: Synthesis of oligoclonal virus-specific IgG antibodies and homogenous free light chains in the central nervous system. *Eur. Neurol.*, **17**, 13–22
163. Townsend, J. J., Wolinsky, J. S. and Baringer, J. R. (1976). The neuropathology of progressive rubella panencephalitis. *Brain*, **99**, 81–90
164. Townsend, J. J., Stroop, W. G., Baringer, J. R., Wolinksy, J. S., McKerrow, J. H. and Berg, B. O. (1982). Neuropathology of progressive rubella panencephalitis after childhood rubella. *Neurology*, **32**, 185–90
165. Rorke, L. B. and Spiro, A. J. (1967). Cerebral lesions in congenital rubella syndrome. *J. Pediatr.*, **70**, 243–55

11
Experimental Mumps Virus Infections in the Developing Nervous System

K. KRISTENSSON, A. LÖVE AND E. NORRBY

INTRODUCTION

Mumps virus often spreads to the human central nervous system and about half of all infections are associated with pleocytosis in the cerebrospinal fluid. The symptoms are usually related to meningeal infections, but signs of encephalitis are not rare[1-3]. The few fatal cases reported seem mainly to represent post-infectious, demyelinating encephalitis[4].

Although mumps virus encephalitis generally is regarded as a mild, transient infection this may not always be the case. Recently, Julkunen et al.[5] in a long-term follow-up study reported that 23 out of 47 patients examined 1-15 years after acute encephalitis associated with mumps virus infection experienced clinical sequelae such as difficulties in memory and learning, focal motor or sensory signs, loss of hearing and visual acuity. The frequency of sequelae in this long-term follow-up study seems to be higher than in previous more short-term follow-up studies[6-8]. This may result from differences in diagnostic criteria of either the acute mumps encephalitis or the follow-up examinations, or indicate that some patients develop a persistent, ongoing process in the brain. The latter situation may have prevailed in two patients who experienced visual deterioration several months after the acute disease and in one patient who developed a slowly progressive encephalomyelitis 14 years after the infection. In cerebrospinal fluids from six of eight examined patients enzyme-linked immunoassay showed antibodies to the IgG isotype and the serum/CSF antibody index was markedly lowered in three cases indicating a persistent intrathecal antibody production[5] against mumps virus.

In a series of experimental studies on the pathogenesis of persistent RNA virus infections in the brain[9] we have investigated interactions of mumps virus and the nervous system. The goal is to evaluate to what extent mild persistent or transient virus infections can cause disturbances in neurotransmission and if non-cytolytic infections during critical time periods in the postnatal development can cause maturation disturbances in the nervous system. Previously, experimental models for mumps CNS infections have been developed[10, 11]. Certain strains of mumps virus can induce an acute fatal encephalitis while others may mainly attack ependymal cells and cause stenosis of the aqueduct with hydrocephalus as a late result[12-14]. Persistent infections in the brain have also been described[15, 16] and a role for viral neuraminidase in neurovirulence was suggested[17, 18].

Here we present some recent data indicating:

(1) defects in expression of certain viral structural proteins in non-cytolytic infections:

(2) pathogenetic determinants in the two viral envelope glycoproteins – the haemagglutinin-neuraminidase (HN) protein and the fusion (F) protein;

(3) disturbances in the development of rod cells in the hamster retina caused by one mutant virus strain;

(4) a peculiar selective neuronal attack by another mutant virus.

NEURONAL RESTRICTION OF VIRUS MULTIPLICATION

In the first series of experiments the appearance of five structural virus proteins in infected neurons *in vivo* and *in vitro* was examined with the aid of monoclonal antibodies against these proteins, namely the HN, F, matrix (M), nucleocapsid (NP) and phospho (P) proteins[19a]. For each structural component two monoclonal antibodies were selected that gave maximal immunofluorescence against mumps virus in infected Vero cells. Newborn hamsters infected intracerebrally with the neuroadapted Kilham virus strain developed an acute fatal encephalitis and high titres of infectious virus were recovered from their brains. All five structural proteins (HN, F, M, NP and P) were detected with immunofluorescence technique in their brains. Newborn mice, on the other hand, showed no signs of disease and no infectious virus was recovered. In spite of this, a large number of neurons were infected as revealed by immunostaining against the NP and P proteins. The antibodies against the HN, F and M proteins gave negative results. In order to study whether this apparent defect in virus replication reflected a restriction exerted by the host cell (the neuron) or by the host animal's defence mechanisms, dissociated hamster and mouse neurons were infected *in vitro*[19b]. Again the hamster neurons showed a productive infection with expression of all antigens. From mouse neuron cultures infectious virus could be recovered only during the first days after infection when there was about 30% loss of neurons. Four days after inoculation NP antigen was present in about 70% of the remaining neurons. There was no expression of the envelope proteins and there was no further loss of neurons. In spite of this, only 10-15% of the neurons contained virus antigens 20 days after inoculation indicating a progressive loss of virus antigens from these non-dividing neurons. These experiments therefore emphasize the important role of the nerve cells' metabolism in restricting viral protein expression.

144

PATHOGENETIC DETERMINANTS OF VIRUS ENVELOPE GLYCOPROTEINS

Previous studies have demonstrated the importance of envelope proteins in the pathogenesis of viral diseases. Most information on such pathogenetic determinants is available from the reovirus model[20], but information is also emerging about other viral systems such as influenza[21], rabies[22,23] and mumps[17], where the neuraminidase activity of the HN protein and the fusion capacity of the F protein[14,18] have been suggested as determinants for neurovirulence. In our studies, on pathogenetic determinants of the virions, we used mutants produced in two different ways. In the first approach, viruses were grown in the presence of a monoclonal antibody directed against a defined epitope of the HN protein in the virus envelope. Since these are neutralizing antibodies only mutant virions with an alteration in the region corresponding to the antibody can be recovered. By this method we isolated one mutant which was found to be much less neurovirulent than the parental Kilham strain[24]. The mutant infected only a few, scattered neurons in the brain, while ependymal cells were heavily infected. This led to the development of hydrocephalus. These experiments therefore show that a small antigenic site of the HN viral protein determines if neurons will be heavily infected with mumps virus or not.

The second method for isolating two mumps virus mutants employed serial passage of viruses recovered from the brain 5 and 10 days after i.c. inoculation of the parental strain, respectively[25]. In this way one strain, rapid Kilham (RK) that caused a rapidly fatal encephalitis in newborn hamsters and another strain, slow Kilham (SK), that caused less acutely lethal disease were isolated. These two mutants displayed several differences not only in their envelope proteins but also in the NP and P proteins as determined by comparative binding to a series of monoclonal antibodies against the structural proteins of mumps virus. In the brain of newborn hamsters the mutants differed in two major respects. The RK strain caused a widespread infection of neurons and a particularly heavy involvement of the caudate nucleus, while the SK strain caused a less extensive neuronal infection with almost no involvement of the caudate nucleus, in contrast to the thalamus and hippocampal regions where a large number of neurons were infected (Figure 11.1). It therefore appears that these two strains differed not only in their degree of neurovirulence but also in their affinity to different regions of the brain. Secondly, the RK strain caused rapid extensive necroses of the brain which were more pronounced than caused by the original strain, while only minimal nerve cell necrosis was found after infection with the SK strain.

In Vero cells the RK strain had a very high fusion capacity while the SK strain caused only minimal cell fusion. A correlation between the fusion capacity and neurovirulence for mumps virus has been demonstrated previously[18,26]. Since cell fusion is induced by the activated viral F protein it was of interest to examine the effect of anti-F monoclonal antibodies on the encephalitis. Two non-neutralizing anti-F antibodies were used and injected subcutaneously at the time of i.c. virus inoculation into the newborn hamsters[27]. Both had a protective effect, which was most

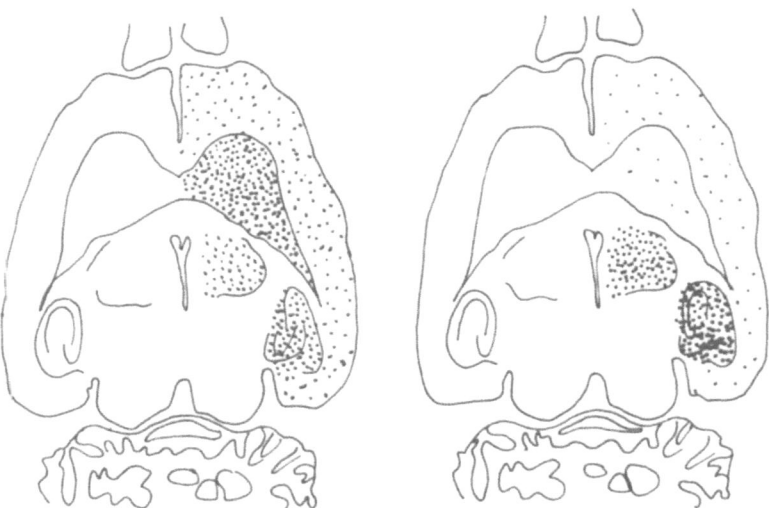

Figure 11.1 Distribution of virus antigen in hamster brain after i.c. inoculation of the RK (left) and SK (right) mumps virus strain. The former causes a heavy infection of the caudate nucleus, while the SK strain mainly infects the hippocampus and thalamic nuclei

pronounced after treatment with one of the antibodies which also inhibited haemolysis *in vitro*. The latter antibody completely protected from the acute fatal disease but, ultimately, the hamsters developed hydrocephalus and died apparently from this. Histologically, there were almost no signs of necrosis in the brain although there was a widespread infection of neurons as revealed by the immunohistochemical technique. This non-neutralizing antibody therefore does not seem to inhibit the infection of the neurons in the brain but to prevent the tissue from undergoing necrosis. This is in contrast to treatment with neutralizing monoclonal antibodies against the HN protein whereby the virus antigen is almost eliminated from the brain[28]. It therefore seems likely that the viral F protein is, at least partially, directly involved in the pathogenesis of cellular destruction in the brain. If the cytolytic action of the viruses is prevented, the basic neuronal structure will remain intact, i.e. the first prerequisite for virus persistence is a neuron. The other prerequisite is, as indicated previously, avoidance of immunosurveillance and a balance in the virus–nerve cell metabolic interaction permitting continuous low grade virus replication.

DEVELOPMENTAL DISTURBANCES OF THE RETINA

The parental Kilham strain of mumps virus regularly spread to the retina after intracerebral inoculation and caused an extensive infection of all retinal layers and widespread necrosis. The mutant strain with an alteration in the HN glycoprotein also spread to the retina and caused a heavy infection of stretches of retinal pigmentary epithelium (RPE), but in the retina only a few neuroectodermal cells were involved and marked necrosis was absent. Instead, two types of developmental disturbances were noted. Firstly absence of photoreceptor segments underneath stretches of RPE loss and, secondly, disordered photoreceptor orientation with fold formation underneath detached RPE and small clusters of pigment-containing cells[29] (Figure 11.2).

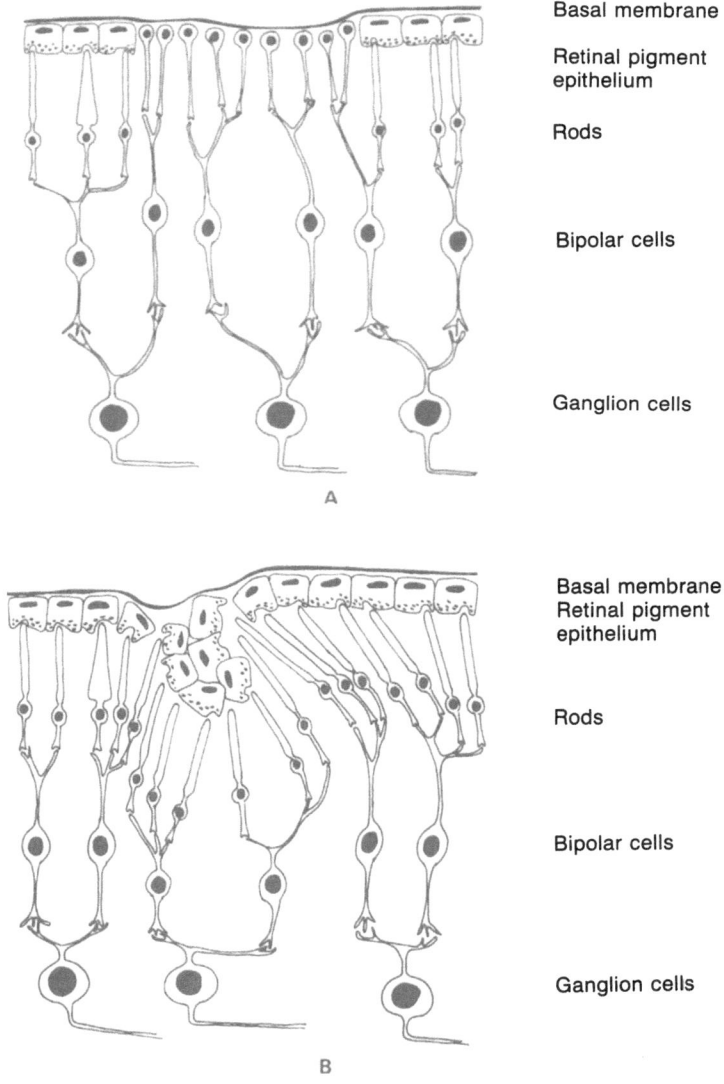

Basal membrane

Retinal pigment
epithelium

Rods

Bipolar cells

Ganglion cells

A

Basal membrane
Retinal pigment
epithelium

Rods

Bipolar cells

Ganglion cells

B

Figure 11.2 Developmental disturbances in the retina after infection with a mumps virus mutant resistant to a monoclonal antibody against the HN-protein: (A) absence of photoreceptor segments underneath loss of infected RPE; (B) disordered photoreceptor orientation underneath detached RPE

Similar retinal dysplastic changes with folds have previously been described after infection with feline leukaemia virus[30] and measles virus[31]. A failure of photoreceptor segment development has been described in immature animals where the RPE has been physically removed[32]. Our finding of a lack of photoreceptor segments in connection with stretches of RPE loss may therefore be explained by a lack of developmental influence from the RPE and may be compatible with the view that the RPE is involved not only in the catabolism of the photoreceptor segments but also in their development. From these experiments it appears that mild infections in the developing nervous system with a mumps virus mutant can cause a disturbance in the orientation of neuroectodermal cells as well as defects in their differentiated development.

CONCLUDING REMARKS

With mutant mumps virus strains alone, or in combination with monoclonal antibodies, tools are now available to induce extensive, non-cytolytic infections both *in vitro* and *in vivo* in such well defined areas of the nervous system as the retina. We have presently also started to examine virus infections in the autonomic ganglia, which represent restricted regions well defined both as to their neurotransmission and synaptogenesis. This will enable us to study more precisely the effects of persistent or defective paramyxovirus infections on the postnatal maturation of neurons, their synapse formation and neurotransmission. The indication obtained from our *in vitro* studies that during certain circumstances a neuron may cure itself from an infection will also be a high priority target for further studies. For instance, can such an infection of immature neurons during a critical stage of their development cause permanent changes in their differentiation, function or life span? If this is the case we will be met with the intriguing or enigmatic situation of a dysfunctioning neuron, which no longer harbours any trace of the causative agent; to clarify the aetiology of such neuronal dysfunctions would then be a major challenge.

ACKNOWLEDGEMENTS

The authors wish to thank Inga-Lisa Wallgren for typing the manuscript. This study was supported by a grant from the Swedish Medical Research Council (project No. B87–12X–04480).

References

1. Scheid, W. (1961). Mumps virus and the central nervous system. *World Neurol.*, **2**, 117–30
2. Bjorvatn, B. and Wolontis, S. (1973). Mumps meningoencephalitis in Stockholm November 1964–July 1971. *Scand. J. Infect. Dis.*, **5**, 253–60
3. Wolinsky, J. S. and Server, A. C. (1985). Mumps virus. In Fields, B. N. (ed.) *Virology*. pp. 1255–84. (New York: Raven Press)
4. Schwarz, G. A., Yang, D. C. and Noone, E. L. (1964). Meningoencephalomyelitis with epidemic parotitis. *Arch. Neurol.*, **11**, 453–62
5. Julkunen, I., Koskiniemi, M., Lehtokoski-Lehtiniemi, E., Saino, K. and Vaheri, A. (1985). Chronic mumps virus encephalitis. *J. Neuroimmunol.*, **8**, 167–75

6. Oldfelt, V. (1949). Sequelae of mumps-meningoencephalitis. *Acta Med. Scand.*, **134**, 405-14

7. Azimi, P. H., Cramblett, H. G. and Haynes, R. E. (1969). Mumps meningoencephalitis in children. *J. Am. Med. Assoc.*, **207**, 509-12

8. Koskiniemi, M., Donner, M. and Pettay, O. (1983). Clinical appearance and outcome in mumps encephalitis in children. *Acta Paediatr. Scand.*, **72**, 603-9

9. Kristensson, K. and Norrby, E. (1986). Persistent RNA viruses in the nervous system. *Annu. Rev. Microbiol.*, **40**, 159-84

10. Overman, J. R., Peers, J. H. and Kilham, L. (1953). Pathology of mumps virus meningoencephalitis in mice and hamsters. *Arch. Pathol.*, **55**, 457-65

11. Johnson, R. T. (1968). Mumps virus encephalitis in the hamster. *J. Neuropathol. Exp. Neurol.*, **27**, 80-95

12. Johnson, R. T. and Johnson, K. P. (1969). Hydrocephalus as a sequela of experimental myxovirus infections. *Exp. Mol. Pathol.*, **10**, 68-80

13. Wolinsky, J. S., Baringer, J. R., Margolis, G. and Kilham, L. (1974). Ultrastructure of mumps virus replication in newborn hamster central nervous system. *Lab. Invest.*, **31**, 403-12

14. McCarthy, M., Jubelt, B., Fay, D. B. and Johnson, R. T. (1980). Comparative studies of five strains of mumps virus *in vitro* and in neonatal hamsters: evaluation of growth, cytopathogenicity, and neurovirulence. *J. Med. Virol.*, **5**, 1-15

15. Wolinsky, J. S., Klassen, T. and Baringer, J. R. (1976). Persistence of neuroadapted mumps virus in brains of newborn hamsters after intraperitoneal inoculation. *J. Infect. Dis.*, **133**, 260-7

16. Wolinsky, J. S. and Stroop, W. G. (1978). Virulence and persistence of three prototype strains of mumps virus in newborn hamsters. *Arch. Virol.*, **57**, 355-9

17. Merz, D. C. and Wolinsky, J. S. (1981). Biochemical features of mumps virus neuraminidases and their relationship with pathogenicity. *Virology*, **114**, 218-27

18. Merz, D. C. and Wolinsky, J. S. (1983). Conversion of non-fusing mumps virus infections to fusing infections by selective proteolysis of the HN glycoprotein. *Virology*, **131**, 328-40

19a. Kristensson, K., Örvell, C., Malm, G. and Norrby, E. (1984). Mumps virus infection of the developing mouse brain – appearance of structural virus proteins demonstrated with monoclonal antibodies. *J. Neuropathol. Exp. Neurol.*, **43**, 131-40

19b. Löve, A., Andersson, T., Norrby, E. and Kristensson, K. (1987). Mumps virus infection of dissociated rodent spinal ganglia *in vitro*. Expression and disappearance of viral structural proteins in neurons. *J. Gen Virol.*, **68**, 1755-9

20. Fields, B. N. and Greene, M. I. (1982). Genetic and molecular mechanisms of viral pathogenesis: implications for prevention and treatment. *Nature*, **300**, 19-23

21. Klenk, H. D., Garten, W., Bosch, F. X. and Rott, R. (1982). Viral glycoproteins as determinants of pathogenicity. *Med. Microbiol. Immunol.*, **170**, 145-53

22. Coulon, P., Rollin, P., Aubert, M. and Flamand, A. (1982). Molecular basis of rabies virus virulence. I. Selection of avirulent mutants of the CVS strain with anti-G monoclonal antibodies. *J. Gen. Virol.*, **61**, 97-100

23. Lafton, M., Wiktor, T. J. and MacFarlan, R. J. (1983). Antigenic sites on the CVS rabies virus glycoprotein: analysis with monoclonal antibodies. *J. Gen. Virol.*, **64**, 843-51

24. Löve, A., Rydbeck, R., Kristensson, K., Örvell, C. and Norrby, E. (1985). Hemagglutinin-neuraminidase glycoprotein as a determinant of pathogenicity in mumps virus hamster encephalitis: analysis of mutants selected with monoclonal antibodies. *J. Virol.*, **53**, 67-74

25. Löve, A., Rydbeck, R., Ljungdahl, Å., Kristensson, K. and Norrby, E. (1986). Selection of mutants of mumps virus with altered structure and pathogenicity by passage *in vivo*. *Microbial. Pathogen.*, **1**, 149-58

26. Merz, D. C., Server, A. C., Waxham, N. N. and Wolinsky, J. S. (1983). Biosynthesis of mumps virus F glycoprotein: non-fusing strains efficiently cleave the F glycoprotein precursor. *J. Gen. Virol.*, **64**, 1957-67

27. Löve, A., Rydbeck, R., Utter, G., Örvell, C., Kristensson, K. and Norrby, E. (1986). Monoclonal antibodies against the fusion protein are protective in necrotizing mumps meningoencephalitis, *J. Virol.*, **58**, 200-2

28. Wolinsky, J. S., Waxham, M. W. and Server, A. C. (1985). Protective effects of glycoprotein-specific monoclonal antibodies on the course of experimental mumps virus meningoencephalitis. *J. Virol.*, **53**, 727-34

29. Löve, A., Malm, G., Rydbeck, R., Norrby, E. and Kristensson, K. (1985). Developmental disturbances in the hamster retina caused by a mutant of mumps virus. *Dev. Neurosci.*, **7,** 65–72

30. Albert, D. M., Lahav, M., Colby, E. D., Shadduck, J. A. and Sang, D. N. (1977). Retinal neoplasia and dysplasia. I. Induction by feline leukemia virus. *Invest. Ophthalmol.*, **16,** 325–37

31. Parhad, I. M., Johnson, K. P., Wolinsky, J. S. and Swoveland, P. (1980). Measles retinopathy. A hamster model for acute and chronic lesions. *Lab. Invest.*, **43,** 52–60

32. Hollyfield, J. and Witkovsky, P. (1974). Pigment retinal epithelium involvement in photoreceptor development and function. *J. Exp. Zool.*, **189,** 359–77

12
Neurotransmitter-Related Activities of MHV3-Infected Cortical Cells in Culture

M. TARDIEU, O. BOESPFLUG AND C. GODFRAIND

INTRODUCTION

The penetration of a virus into the developing nervous system could lead to an acute destruction of the target cells of the virus[1] or to a low grade infection with viral persistence in different types of CNS cells. A classical example of the latter situation is the ability of the rubella virus to persist within human brain cells for several months after the fetal infection.

An important question then, is the understanding of the effect of a non-cytopathogenic viral infection on the different functions of chronically infected CNS cells. Thus, some evidence has suggested that the ability of CNS cells to proliferate was modified after persistent fetal rubella infection. The persistent infection might also modify other metabolic functions of virus-containing neurons and astrocytes and more specifically could alter the neurotransmitter activities of infected CNS cells.

The aim of our work was to study the neuromodulating activities of virus-replicating neurons and astrocytes. To approach this problem, we used the chronic CNS infection induced in C3H mice by a coronavirus, the MHV3, as an experimental model.

IN VIVO INDUCTION OF THE PERSISTENT MHV3 INFECTION OF THE CNS AND ITS NEUROPATHOLOGICAL CONSEQUENCES

Intraperitoneal injection of MHV3 into adult C3H mice results either in early death due to hepatitis or in the development of a chronic disease with neurological manifestations and virus persistence in the surviving animals, The balance between acute hepatitis and chronic CNS disease depends both upon the age of recipient mice and the injected dose of virus (injection of 1 to 5×10^2 PFU of MHV3 into 12–14-week-old mice seems optimal for obtaining the highest percentage of chronic disease with minimal acute death[2]). Chronic neurological disease results from meningitis, ependymitis and encephalitis, beginning 3 to 4 weeks after infection and followed by a permanent communicating hydrocephalus. Later (after the sixth week post infection), a chronic thrombotic vasculitis develops affecting meningeal and parenchymal vessels at the brain stem level. There is no evidence of alteration to white matter, suggesting that *in vivo* MHV3 induces lesions predominantly in meningeal cells, ependymal cells and neurons[3-5].

Some strains of mice are resistant to MHV3-induced diseases (A/Jx mice) while others are highly susceptible (C_{57}/Bl_6 mice) and develop an acute hepatic necrosis leading to death within a few days in all injected animals. The presence of MHV3 can be demonstrated in the liver of A/Jx mice in the days following viral inoculation. However, in this strain of mice, as compared to C_{57}/Bl_6, a higher titre of virus is required for CNS penetration. In spite of viral invasion, no neuropathological lesions are observed in the brain of infected A/Jx mice[2,3].

SELECTIVE AFFINITY *IN VITRO* OF PURIFIED MHV3 FOR EPENDYMAL CELLS, NEURONS AND MENINGEAL CELLS

Purified MHV3 is obtained by polyethylene glycol precipitation of proteins from supernatants of infected L cell cultures and viral isolation on a sucrose gradient (titres between 5×10^6 and $5 \times 10^7 \, PFU \, ml^{-1}$).

Ependymal cells and oligodendrocytes

Cells, isolated by a selective release from mouse brain slices (either with EGTA or DNAase) are purified on a BSA gradient[2]. Purified MHV3 bind to 82% of isolated ciliated ependymal cells (Figure 12.1) but only to 25% of cells present in the oligodendrocyte-enriched cell suspension, as judged by indirect immunofluorescence. In double labelling experiments with anti-galactocerebroside (a marker for oligodendrocytes) and anti-MHV3 antibodies, 23% of the anti-galactocerebroside positive cells are double labelled by anti-MHV3 antibodies. Thus, MHV3 appears to have a stronger affinity for ependymal cells than for oligodendrocytes.

Neurons and astrocytes

Fetal or neonatal dissociated cortical brain cultures provide monolayers of mixed neurons and astrocytes or of astrocytes without neurons, respectively. In both kinds of culture a small percentage of cells (5–10%) are fibroblasts. Neurons and astrocytes are recognized by their morphological aspect under phase microscope and by specific antigenic markers such as GFAP for astrocytes of N-Cam and A2 B5 antigens for neurons. Fetal cortical neurons are brightly stained by indirect immunofluorescence using purified MHV3 and corresponding antibodies. Viral binding is observed on the surface of the majority of examined neurons, on cell bodies as well as on neurites. Neurons of different ages (3–20 days in culture) are equally stained. In sharp contrast, the non-neuronal cells are not stained during the procedure[2].

When similar cultures are infected with 10^2 PFU of MHV3 and are tested 24 or 48 h later for the presence of intracytoplasmic MHV3 antigens, 80 to

85% of neurons are stained, whereas non-neuronal cells are negative. Moreover, MHV3-infected glial cell cultures obtained from newborn mice (containing no neurons) are negative for the presence of intracytoplasmic viral antigens except for 4–6% of cells in cultures tested more than 4 days post infection. Cultured cells were subsequently double-labelled for the presence of intracytoplasmic MHV3 antigen and GFAP, A2 B5 or N-Cam antigen: no GFAP positive cells are positive for MHV3 whereas more than 80% of the A2 B5 and N-Cam positive cells are also MHV3 antigen-positive (Figure 12.1).

Meningeal cells

Cultures of meningeal cells were obtained from explants of fetal meninges. Intracytoplasmic MHV3 antigens are easily detected by indirect immuno-fluorescence in the cytoplasm of meningeal cells in culture, 24 h after MHV3 infection.

CONSEQUENCES OF MHV3 INFECTION ON MORPHOLOGICAL ASPECTS AND FUNCTIONS OF CULTURED CNS CELLS

To further analyse the consequences of MHV3 infection on the metabolism of the host cell, cultures of meningeal cells, astrocytes and neurons have been infected with different doses of MHV3 and several parameters have been monitored during 14 days post infection.

Survival of infected cells and induction of cellular fusion

Meningeal cells and L cells

When infected with 10^2 PFU of MHV3, meningeal cells as well as L cells form large syncytia after 12 h and are disrupted within 48 h. Viral titres in supernatants of infected L cells and meningeal cells increase at a similar rate (between 5×10^4 and 5×10^5 PFU ml^{-1} at day 2 post infection).

Neurons and astrocytes

Morphological maturation of cells in virus-infected (10^2 PFU per culture) and non-infected cortical neuronal cell cultures is similar for 6 to 7 days post infection, after which neurons appeared progressively disrupted over a 3-day period, whereas the morphological aspect of astrocytes is preserved. Formation of neuronal syncytium was occasionally observed on 1 μm sections. The morphological aspect of astrocytes from glial cell cultures is identical in infected (10^2 PFU per culture) and uninfected cultures. Some astrocyte disruption without fusions were, however, observed[2] when cultures were infected with more than 10^4 PFU of virus per culture. (Multiplicity of infection (MOI) > 0.01.)

To further analyse the respective susceptibilities of astrocytes and neurons to MHV3, two continuous cell lines, one of glial origin (C6 glioma) and the other of neuronal lineage (neuroblastoma NIE 115), were tested.

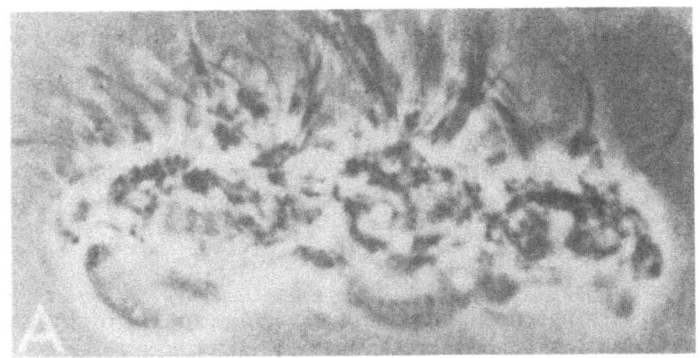

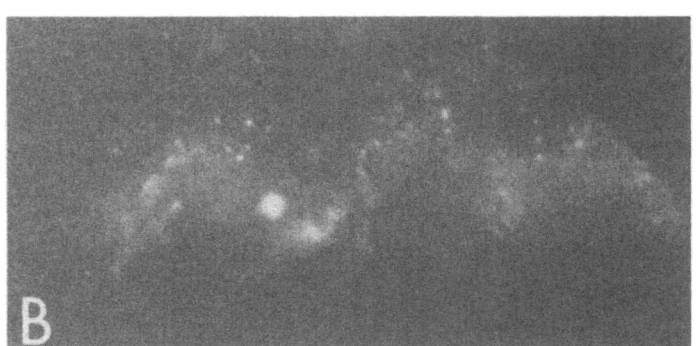

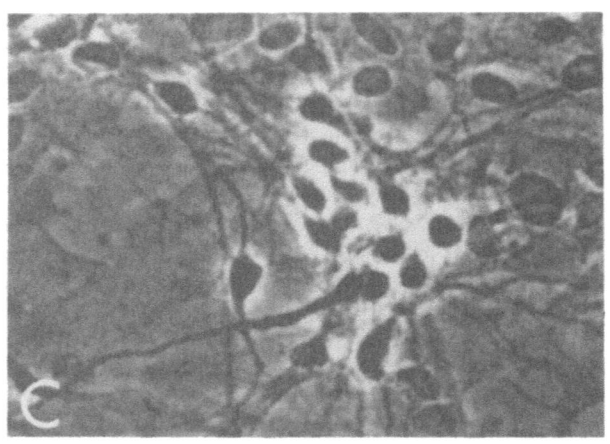

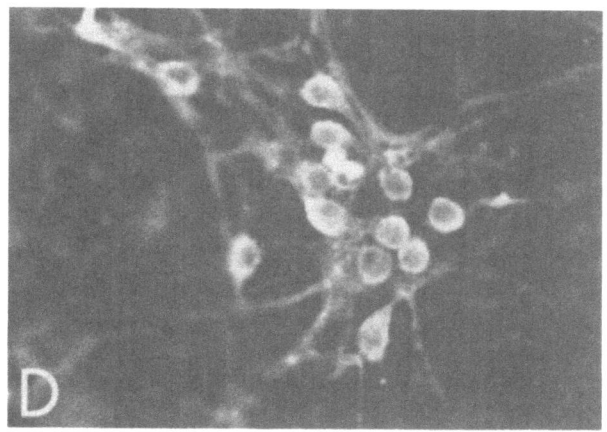

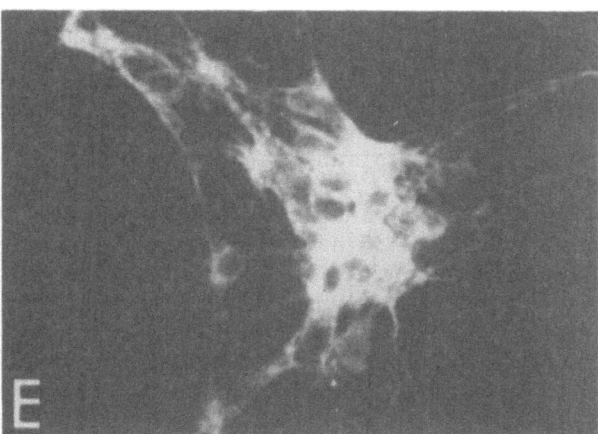

Figure 12.1 (Left panel) Binding of purified MHV3 to isolated ependymal cells of adult C3H mice: (A) isolated ependymal cells as seen under phase microscope; (B) binding of MHV3 as shown by direct immunofluorescence (× 480). (Right panel) MHV3-infected fetal cortical cell culture: (C) as seen under phase microscope 48 h after infection; double-labelled (D) for MHV3 antigen and (E) for N-Cam antigen (× 280)

Neuroblastoma cells when infected at a MOI of 10^{-4}, were destroyed within 36 h without syncytia formation. In sharp contrast, C6 glial cells appeared unaffected by MHV3 infection.

Viral titres in supernatants of 3–6-days-old neuronal cultures or of neuroblastoma cell cultures increased sharply during the first days of infection. Conversely, no virus was found in the supernatant of infected C6 glioma cells whereas viral titres increased slowly in supernatants of glial cell cultures. At days 1 and 2 post infection viral titres in the supernatant of infected glial cells were significantly lower than those observed in super-natants of infected neuronal cultures. By the 3rd day post infection viral titres in supernatants of both types of cultures were identical[2].

Metabolic activities of MHV3-infected neurons and astrocytes

The selective infection of cultured neurons with a relative preservation of astrocytes is of particular interest since a destruction of astrocytes induces by itself a disruption of neuronal cells. This, associated with the delayed expression of viral-induced neuronal lesions, provides a privileged approach to study metabolic activities, and more specifically neurotransmitter-related activities, of virus-infected neurons and astrocytes. Results of experiments in progress will be briefly summarized.

Neurons

In fetal cortical neuronal culture, the protein content per well decreased in infected cultures only after the 7th day post infection, reflecting neuronal death, whereas thymidine incorporation reflecting cellular proliferation was unaffected. The activities of two intra-neuronal enzymes related to neuro-transmitters synthesis (glutamic acid decarboxylase for γ-aminobutyric acid synthesis and choline acetyltransferase for acetylcholine synthesis) were similar in control and MHV3 infected cultures up to the 7th day post infection, i.e. 3–4 days after the peak of viral replication. After the 7th day post infection, they decreased in infected cultures, again reflecting neuronal death. The membrane binding of a neuron-specific benzodiazepine (BDZ) ($[^3H]$methyl-clonazepam) was preserved until the 10th day post infection. In sharp contrast, the high affinity uptake of $[^3H]\gamma$-aminobutyric acid did not increase in infected cultures as it did in the control during neuronal matura-tion. This uptake was significantly lower in infected cultures as compared to the control after the 4th day post infection. Kinetic curves performed at day 6 post infection, suggested a selective decrease of the maximum velocity with no alteration of the affinity.

Astrocytes

In cortical glial cell culture, the protein content was similar in virus infected (10^2 PFU per culture) and control cultures whereas cellular proliferation, judged by thymidine incorporation was slightly reduced after the sixth day post infection. The activity of glutamine synthetase (a glial-specific enzyme) was unaffected by viral infection unless very high doses of virus were used. The binding of the non-neuronal cell-specific benzodiazepine, the Ro5-4864 decreased after the fourth day post infection; this effect was dose-dependent. It also depended on the ability of the virus to replicate: if 10^4 PFU of virus were UV-irradiated the effect was abolished, suggesting that the binding of viral particles to the cell surface was not itself directly responsible for the decreased BDZ binding. A Scatchard plot at day 7 post infection showed a decreased affinity of the high affinity BDZ receptor in the infected culture, whereas the maximum binding was the same in both types of cultures. No modification of the low affinity BDZ receptor was observed.

DISCUSSION

Purified MHV3 has a high affinity for neurons, ependymal cells and meningeal cells *in vitro* but not for astrocytes and oligodendrocytes. This corresponds to its pattern of pathogenicity *in vivo* since MHV3 induces an initial ependymitis, meningitis and encephalitis in the absence of any white matter lesion. The difference in susceptibility to MHV3 between neurons and astrocytes is specially striking: neurons but not astrocytes appear to be slowly destroyed after viral infection while viral replication is very active in the supernatant of young embryonic cultures when neuronal cells predominate. Infected glial cells in culture, however, retain some susceptibility to MHV3 since evidence of slow viral replication is observed. This pattern of affinity between CNS cells and MHV3 differs from that observed with other corona-viruses[6-9]. Restricted tropism of mouse hepatitis viruses for CNS cells could be due to an antigenic variation in the viral attachment glycoprotein, E_2[10].

In vitro metabolic activity of MHV3-infected maturing neurons can be preserved for several days even though neurons are actively replicating the virus. Thus, the intraneuronal neurotransmitter-related enzymatic activities and the affinity of membrane receptors for a specific ligand (BDZ) are identical in infected and control cultures whereas the high affinity membrane γ-aminobutyric acid uptake is altered but not abolished. Similarly, MHV3 infection induces an alteration of astrocyte proliferation and of the affinity of membrane receptors for BDZ without modification of an intracellular enzymatic activity. These results show that MHV3 replication, when it does not induce an acute cytopathogenic effect, could alter the 'luxury functions' of CNS host cells in a very complex way.

More generally, when a virus induces a non-cytopathogenic infection, as in the case of several fetal infections, this chronic infection might alter neurotransmitter activities of infected fetal neurons. These alterations could react upon fetal brain either directly, modifying individual cell functions, or indirectly, altering cellular interconnection in the maturing central nervous system.

Acknowledgements

This work was supported by INSERM grant CRL 82 60 36 and by grant no. 884 from Université Paris Sud. Catherine Godfraind is supported by FNRS, Université Catholique de Louvain, Belgium.

References

1. Tardieu, M., Epstein, R. L. and Weiner, H. L. (1982). Interactions of viruses with cell surface receptors. *Int. Rev. Cytol.*, **80**, 27-61
2. Tardieu, M., Boespflug, O. and Barbé, T. (1986). Selective tropism of a neurotropic coronavirus for ependymal cells, neurons and meningeal cells. *J. Virol.*, **60**, 574-82
3. Tardieu, M., Goffinet, A., Harmand-Van Rijckevorsel, G. and Lyon, G. (1982). Ependymitis, hydrocephalus and vasculitis following chronic infection by mouse hepatitis virus 3 (MHV3): role of genetic and immunological factors. *Acta Neuropathol. (Berl.)*, **58**, 168-76
4. Virelizier, J. L., Dayan, A. D., and Allison, A. C. (1975). Neuropathologic effects of persistent infection of mice by mouse hepatitis virus, *Infect. Immun.*, **12**, 1127-40
5. Le Prevost, C., Virelizier, J. L. and Dupuy, J. M. (1975). Immunopathology of mouse hepatitis type 3 infection. III. Clinical and virologic observation of a persistent viral infection. *J. Immunol.*, **115**, 640-3
6. Dubois-Dalcq, M. E., Doller, E. W., Haspel, M. V. and Holmes, K. V. (1982). Cell tropism and expression of mouse hepatitis viruses (MHV) in mouse spinal cord cultures. *Virology*, **119**, 317-31
7. Knobler, R. L., Haspel, M. V. and Oldstone, M. B. A. (1981). Mouse hepatitis virus type 4 (JHM strain)-induced fatal central nervous system disease. I. Genetic control and the murine neuron as the susceptible site of disease. *J. Exp. Med.*, **153**, 832-43
8. Collins, A. R., Tunison, L. A. and Knobler, R. L. (1983). Mouse hepatitis virus type 4 infection of primary glial cultures from genetically susceptible and resistant mice. *Infect. Immun.*, **40**, 1192-7
9. Suzumar, A., Lavi, E., Weiss, S. R. and Silberberg, D. H. (1986). Coronavirus infection induces H2 antigen expression on oligodendrocytes and astrocytes. *Science*, **232**, 991-3
10. Fleming, J. O., Stohlman, S. A., Harmon, R. C., Lai, M. M. C., Frelinger, J. A. and Weiner, L. P. (1983). Antigenic relationships of murine coronaviruses: Analyses using monoclonal antibodies to JHM (MHV4) virus. *Virology*, **131**, 296-307

Index